HEINEMANN COORDINATED SCIENCE • HIGHER

PHYSICS

Graham Dolan • Mike Duffy • Adrian Percival

Heinemann

Contents

How to use this book   1

SECTION A: ELECTRICITY AND MAGNETISM

CHAPTER 1: WHAT IS ELECTRICITY?

1.1 *Electric charge*	2	
1.2 *Hazards and uses of static electricity*	4	
1.3 *Electric current, voltage and circuit diagrams*	6	
1.4 *Resistance and Ohm's Law*	8	
1.5 *I–V relationships*	10	
1.6 *Series and parallel circuits*	12	
1.7 *Movement of charge*	14	
1.8 *Electrolysis; cathode ray oscilloscope*	16	
1.9 *Electrical power*	18	

CHAPTER 2: ELECTRICITY IN ACTION

2.1 *Mains supply*   20
2.2 *Household wiring*   22
2.3 *Domestic appliances*   24
2.4 *The cost of electricity*   26

CHAPTER 3: USING MAGNETS

3.1 *Magnetism and electricity*   28
3.2 *Electromagnets*   30
3.3 *Electric motors*   32
3.4 *Electromagnetic induction*   34
3.5 *Transformers*   36
3.6 *Generation and transmission of electricity*   38
Section A: Questions   40

SECTION B: FORCES AND MOTION

CHAPTER 4: FORCES AND MOVEMENT

4.1 *Force*   42
4.2 *Resultant force*   44
4.3 *Speed*   46
4.4 *Acceleration*   48
4.5 *Force and acceleration*   50
4.6 *Friction*   52
4.7 *Turning effects*   54

CHAPTER 5: FORCES, SHAPE AND PRESSURE

5.1 *Solids, liquids and gases*   56
5.2 *Elasticity*   58
5.3 *Density*   60
5.4 *Pressure*   62
5.5 *Hydraulic systems*   64
5.6 *Pressure in gases*   66
Section B: Questions   68

SECTION C: WAVES

CHAPTER 6: CHARACTERISTICS OF WAVES

6.1 *What are waves?*   70
6.2 *Looking at waves in a ripple tank*   72
6.3 *Electromagnetic waves*   74
6.4 *Infra-red and ultra-violet*   76
6.5 *X-rays and gamma rays*   78
6.6 *Light is a wave*   80
6.7 *Refraction of light*   82
6.8 *Total internal reflection*   84
6.9 *Diffraction*   86

CHAPTER 7: SOUND

7.1 *How are sounds produced?*	88	**7.4** *Ultrasound* 94
7.2 *The speed of sound*	90	**7.5** *Seismic waves* 96
7.3 *Picturing sound waves*	92	**Section C: Questions** 98

SECTION D: ENERGY RESOURCES AND ENERGY TRANSFER

CHAPTER 8: MOVING ENERGY AROUND

8.1 *Where does energy come from?*	100	**8.5** *Radiation* 108
8.2 *Energy conversions*	102	**8.6** *Saving energy at home* 110
8.3 *Transferring thermal energy*	104	**8.7** *Potential and kinetic energy* 112
8.4 *More about conduction and convection*	106	**8.8** *Work and power* 114

CHAPTER 9: ENERGY SUPPLIES

9.1 *Using energy*	116	**9.5** *Using solar energy* 124
9.2 *Power stations*	118	**9.6** *Energy and money* 126
9.3 *Using water to generate electricity*	120	**Section D: Questions** 128
9.4 *Energy from wind, hot rocks and waste*	122	

SECTION E: RADIOACTIVITY

CHAPTER 10: RADIOACTIVITY

10.1 *What is radioactivity?*	130	**10.5** *Radioactivity and you* 138
10.2 *Properties and detection of nuclear radiation*	132	**10.6** *Radiation at work* 140
10.3 *Radioactive decay*	134	**10.7** *The source of nuclear energy* 142
10.4 *Radioactivity and dating*	136	**Section E: Questions** 144

SECTION F: EARTH IN SPACE

CHAPTER 11: THE EARTH AND BEYOND

11.1 *Our solar syatem*	146	**11.6** *The evolution of stars* 156
11.2 *The planets*	148	**11.7** *The galaxies* 158
11.3 *Periodic changes*	150	**11.8** *The evolution of our solar syatem* 160
11.4 *What revolves around what?*	152	**11.9** *Moving round in circles* 162
11.5 *Stars and energy*	154	**Section F: Questions** 164

Index   166

Heinemann Educational Publishers
Halley Court, Jordan Hill, Oxford, OX2 8EJ
a division of Reed Educational & Professional Publishing Ltd

OXFORD FLORENCE PRAGUE MADRID ATHENS
MELBOURNE AUCKLAND KUALA LUMPUR
SINGAPORE TOKYO IBADAN NAIROBI KAMPALA
JOHANNESBURG GABORONE PORTSMOUTH NH
CHICAGO MEXICO CITY SAO PAULO

© Graham Dolan, Mike Duffy and Adrian Percival, 1996

Copyright notice

All rights reserved. No part of this publication may be reproduced, stored in a retrieval system, or transmitted in any form or by any means, electronic, mechanical, photocopying, recording, or otherwise without either the prior written permission of the Publishers or a licence permitting restricted copying in the United Kingdom issued by the Copyright Licensing Agency Ltd, 90 Tottenham Court Road, London W1P 9HE.

First published 1996

ISBN 0 435 58001 9

2000 99 98 97 96
10 9 8 7 6 5 4 3 2 1

Designed and typeset by Ken Vail Graphic Design

Illustrated by: Simon Girling and Associates (Mike Lacey, Catherine Ward), Nick Hawken, John Plumb, Ken Vail Graphic Design.

Cover design by Ken Vail Graphic Design

Cover photo by SPL/P. Nunuk (inset: Bruce Coleman/NASA)

Printed and bound in Spain by Mateu Cromo

Acknowledgements

The authors and publishers would like to thank the following for permission to use photographs:

p 2 *T*: J. Allan Cash Ltd. **p 2** *MR*: Trevor Hill. **p 2** *ML*: Image Select/Ann Ronan. **p 4** *T*: Frank Lane Picture Agency. **p 4** *M*: SPL/Phil Jude. **p 5** *B*: Environmental Picture Library/Philip Carr. **p 6** *T*: Image Select/Ann Ronan. **p 7** *all*: Peter Gould. **p 8** *MR*: Peter Gould. **p 8** *BL*: Image Select/Ann Ronan. **p 14** *M*: Image Select/Ann Ronan. **p 16** *M*: J.C. Davies Photography, Holyhead. **p 19** *T*: Peter Gould. **p 20** *T (both)*: Peter Gould. **p 21** *both*: Peter Gould. **p 22** *M*: Peter Gould. **p 23** *both*: Peter Gould. **p 24** *M (both)*: Peter Gould. **p 25** *T*: Peter Gould. **p 26** *M (both)*: Peter Gould. **p 28** *BR*: SPL/Vaughan Fleming. **p 29** *TL*: Image Select/Ann Ronan. **p 29** *TR*: Peter Gould. **p 30** *TR*: SPL/Alex Bartel. **p 30** *ML*: Zefa. **p 31** *MR*: SPL/David Parker/600 Group Fanue. **p 32** *T*: SPL/David Ducros/Jerrican. **p 36** *T*: Peter Gould. **p 37** *M*: Peter Gould. **p 38** *T*: The Glasgow Picture Library. **p 38** *M*: SPL/James Stevenson. **p 42** *T*: Select Image/Ann Ronan. **p 43** *L*: Peter Gould.

p 43 *R*: SPL/Prof. Harold Edgerton. **p 44** *T*: Chris Honeywell. **p 46** *MR*: Chris Honeywell. **p 46** *B*: Quadrant Picture Library. **p 49** *T*: Planet Earth Pictures. **p 49** *M*: Quadrant Picture Library. **p 50** *T*: Planet Earth Pictures. **p 50** *B*: Peter Gould. **p 51**: J. Allen Cash Ltd. **p 52** *T*: Allsport/Pascal Rondeau. **p 52** *B*: Quadrant Picture Library. **p 53** *L*: NHPA/Gerard Lacz. **p 54** *T*: Graham Dolan. **p 55** *T*: Chris Honeywell. **p 56** *T*: Chris Honeywell. **p 57** *B (both)*: Peter Gould. **p 60** *B*: Peter Gould. **p 61** *T*: Peter Gould. **p 62** *T, M & B*: Chris Honeywell. **p 63** *T*: Image Select/Ann Ronan. **p 63** *ML*: Silvestris/Frank Lane Picture Agency/D Fleetham. **p 64** *T (both)*: J. Allan Cash Ltd. **p 66** *T*: Chris Honeywell. **p 67** *T*: Image Select/Ann Ronan. **p 70** *T*: J. Allan Cash Ltd. **p 73** *T*: J. Allan Cash Ltd. **p 75** *MR*: Topham Picturepoint. **p 76** *T*: Shout Picture Library. **p 77** *T*: J. Allan Cash Ltd. **p 78** *T*: Courtesy of The Nebel Collection, Image Select/Ann Ronan(Custom Medical Stock Photo). **p 78** *MR*: SPL/Samuel Giannavola. **p 78** *B*: The Wellcome Institute Library, London. **p 79** *T*: Trevor Hill. **p 81** *M*: Trevor Hill. **p 83** *B*: Topham Picturepoint. **p 84** *T*: NHPA. **p 85** *T*: Science Photo Library/Will & Deni McIntyre. **p 85** *ML*: SPL/BSIPB Bajande. **p 86** *M*: Sealand Aerial Photography. **p 86** *BR*: Peter Gould. **p 88** *T*: EBET Roberts Redferns. **p 88** *B*: NHPA/Andy Rouse. **p 92** *M*: Peter Gould. **p 92** *B*: Peter Gould. **p 93** *T*: Peter Gould. **p 93** *M*: Topham Picturepoint/R. Setford. **p 94** *T*: NHPA. **p 94** *BR*: Marconi. **p 96** *T*: S. Mikami/Press Association/Topham Picturepoint. **p 100** *T*: Image Select/Ann Ronan. **p 102** *B*: Peter Gould. **p 103** *T*: Peter Gould. **p 107** *T*: Topham Picturepoint. **p 109** *T*: Zefa Pictures. **p 109** *M*: Chris Honeywell. **p 110** *B*: SPL/Agema Infrared. **p 111**: Peter Gould. **p 117** *B*: Frank Spooner Pictures/Bernstein/Sponner. **p 118** *T*: SPL/Roger Ressmeyer, Starlight. **p 119** *T*: SPL/Novosti **p 119** *B*: London Aerial Photo Library. **p 120** *T*: SPL/John Mead. **p 120** *B*: Topham Picturepoint. **p 121**: News Photo Library/Press Association. **p 122** *M*: The Glasgow Picture Library. **p 123** *T*: SPL/Simon Fraser. **p 124** *T*: Environmental Picture Library. **p 124** *M*: Trevor Hill. **p 124** *B*: SPL/Martin Bond. **p 125** *B*: J. Allan Cash Ltd. **p 127** *M*: National Grid Company. **p 130** *T*: Image Select/Ann Ronan. **p 132** *M*: Peter Gould. **p 133** *BL*: Peter Gould. **p 136** *T*: Radiocarbon Dating. **p 136** *M*: SPL/Gianni Tortoli. **p 136** *B*: Topham Picturepoint. **p 140** *M (both)*: Peter Gould. **p 140** *B*: NRPB. **p 141**: SPL/CNRI. **p 142** *M*: SPL/Les Almes National Laboratory. **p 143** *M*: Environmental Picture Library/Dilan Garcia. **p 143** *B*: SPL/U.S. Department of Energy. **p 148** *B*: SPL/NASA. **p 149** *T*: SPL/NASA. **p 149** *B*: NASA/Science Photo Library. **p 151** *T*: SPL. **p 151** *ML*: SPL/Hale Observatories. **p 152** *ML*: SPL/Dr Jeremy Burgess. **p 153** *T*: SPL/Rev Ronald Royer. **p 153** *BL*: National Maritime Musuem Picture Library. **p 154** *T*: SPL/NOAO. **p 155** *T*: SPL. **p 157** *T*: SPL/Space Telescope Science Institute/NASA. **p 157** *M (both)*: SPL/NOAO. **p 158** *T*: SPL/Royal Observatory, Edinburgh/AATB. **p 158** *M*: Image Select/Ann Ronan. **p 161** *M*: Michael Holford. **p 161** *B*: SPL/Francois Gohier. **p 163**: SPL/GE Astro Space.

p 47: Crown copyright is reproduced with the permission of the Controller of HMSO; **p 116** and **p 117**: graphs from the Department of Trade and Industry; **p 126**: graph from Electricity Association; **p 127**: graph from the National Grid; **p 138**, **p 139** and **p 145**: graphs and map from the National Radiological Protection Board; **p 152**: diagram of path of Mars from Harry Ford.

The publishers have made every effort to trace the copyright holders, but if they have inadvertently overlooked any, they will be pleased to make the necessary arrangements at the first opportunity.

How to use this book

Heinemann Coordinated Science: Physics has been written for your GCSE course and contains all the information you will need over the next two years for your exam syllabus.

This book has six sections. Each section matches one of the major themes in the National Curriculum.

What is in a section?

The sections are organised into double-page spreads. Each spread has:

Colour-coded sections so you can quickly find the one you want.

Clear text and pictures to explain the science.

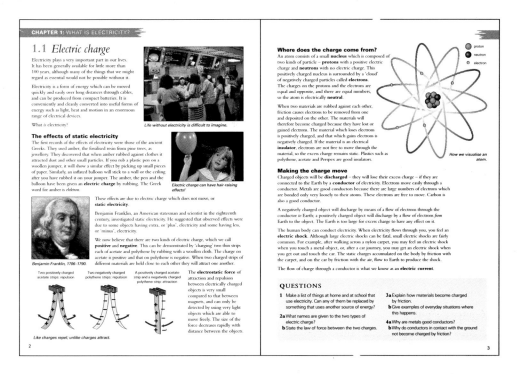

Questions to help check your understanding of the important ideas on the spread.

At the end of each section, there are double-page spreads of longer questions. These are to help you find out if you understand the key ideas in that section. They can also help you revise.

Assessment and resource pack

All the answers for questions in this student book are in the *Heinemann Coordinated Science: Higher Physics Assessment and resource pack*.

CHAPTER 1: WHAT IS ELECTRICITY?

1.1 Electric charge

Electricity plays a very important part in our lives. It has been generally available for little more than 100 years, although many of the things that we might regard as essential would not be possible without it.

Electricity is a form of energy which can be moved quickly and easily over long distances through cables, and can be produced from compact batteries. It is conveniently and cleanly converted into useful forms of energy such as light, heat and motion in an enormous range of electrical devices.

What *is* electricity?

Life without electricity is difficult to imagine.

The effects of static electricity

The first records of the effects of electricity were those of the ancient Greeks. They used amber, the fossilised resin from pine trees, as jewellery. They discovered that when amber rubbed against clothes it attracted dust and other small particles. If you rub a plastic pen on a woollen jumper, it will show a similar effect by picking up small pieces of paper. Similarly, an inflated balloon will stick to a wall or the ceiling after you have rubbed it on your jumper. The amber, the pen and the balloon have been given an **electric charge** by rubbing. The Greek word for amber is *elektron*.

Electric charge can have hair-raising effects!

These effects are due to electric charge which does not move, or **static electricity**.

Benjamin Franklin, an American statesman and scientist in the eighteenth century, investigated static electricity. He suggested that observed effects were due to some objects having extra, or 'plus', electricity and some having less, or 'minus', electricity.

We now believe that there are two kinds of electric charge, which we call **positive** and **negative**. This can be demonstrated by 'charging' two thin strips each of acetate and polythene by rubbing with a woollen cloth. The charge on acetate is positive and that on polythene is negative. When two charged strips of different materials are held close to each other they will attract one another.

Benjamin Franklin, 1706–1790.

Two positively charged acetate strips: repulsion

Two negatively charged polythene strips: repulsion

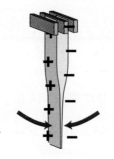
A positively charged acetate strip and a negatively charged polythene strip: attraction

The **electrostatic force** of attraction and repulsion between electrically charged objects is very small compared to that between magnets, and can only be detected by using very light objects which are able to move freely. The size of the force decreases rapidly with distance between the objects.

Like charges repel; unlike charges attract.

Where does the charge come from?

An atom consists of a small **nucleus** which is composed of two kinds of particle – **protons** with a positive electric charge and **neutrons** with no electric charge. This positively charged nucleus is surrounded by a 'cloud' of negatively charged particles called **electrons**. The charges on the protons and the electrons are equal and opposite, and there are equal numbers, so the atom is electrically **neutral**.

When two materials are rubbed against each other, friction causes electrons to be removed from one and deposited on the other. The materials will therefore become charged because they have lost or gained electrons. The material which loses electrons is positively charged, and that which gains electrons is negatively charged. If the material is an electrical **insulator**, electrons are not free to move through the material, so the excess charge remains static. Plastics such as polythene, acetate and Perspex are good insulators.

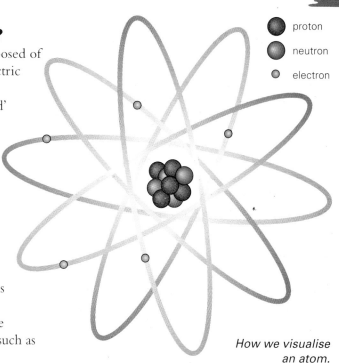

How we visualise an atom.

Making the charge move

Charged objects will be **discharged** – they will lose their excess charge – if they are connected to the Earth by a **conductor** of electricity. Electrons move easily through a conductor. Metals are good conductors because there are large numbers of electrons which are bonded only very loosely to their atoms. These electrons are free to move. Carbon is also a good conductor.

A negatively charged object will discharge by means of a flow of electrons through the conductor *to* Earth; a positively charged object will discharge by a flow of electrons *from* Earth to the object. The Earth is too large for excess charge to have any effect on it.

The human body can conduct electricity. When electricity flows through you, you feel an **electric shock**. Although large electric shocks can be fatal, small electric shocks are fairly common. For example, after walking across a nylon carpet, you may feel an electric shock when you touch a metal object, or, after a car journey, you may get an electric shock when you get out and touch the car. The static charges accumulated on the body by friction with the carpet, and on the car by friction with the air, flow to Earth to produce the shock.

The flow of charge through a conductor is what we know as an **electric current**.

QUESTIONS

1. Make a list of things at home and at school that use electricity. Can any of them be replaced by something that uses another source of energy?

2a. What names are given to the two types of electric charge?
 b. State the law of force between the two charges.

3a. Explain how materials become charged by friction.
 b. Give examples of everyday situations where this happens.

4a. Why are metals good conductors?
 b. Why do conductors in contact with the ground not become charged by friction?

1.2 Hazards and uses of static electricity

Great sparks!

A charged object has **electric potential energy**. We say there is a **potential difference** (or **voltage**) between it and Earth. The greater the concentration of charge, the greater the potential difference. Sparks are caused by a high potential difference – the highly charged object discharges through the air (normally an insulator) and sparks are seen jumping between the object and Earth or a conductor connected to Earth.

Lightning results from a cloud building up enough charge for a spark to jump from cloud to cloud or from the cloud to Earth.

Lightning can cause severe damage and fatal injury when it strikes the ground. High buildings are at greatest risk but damage can be prevented by the use of a lightning conductor. This consists of a strip of copper (a very good conductor) connecting a metal spike above the building to a metal plate in the ground.

Benjamin Franklin flew a kite into thunderclouds to 'collect' some electricity. He wanted to prove that lightning was a huge electric spark. Warning! This experiment was very dangerous.

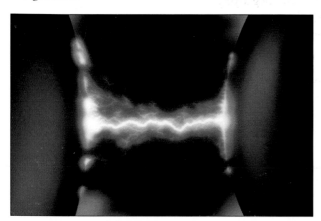

Sometimes enough charge can build up to cause sparks.

When aircraft are being refuelled, the hose could become charged by friction as the fuel flows through it. An electrical connection is made between the fuel tanker, the hose, the aircraft and the Earth before the flow of fuel starts. This prevents sparking which might ignite the flammable fuel vapour with explosive results.

Similar precautions are taken when petrol is delivered to filling stations. Charge builds up on petrol tankers because of friction between the fuel and the tank, and between the tanker and the air.

When an object is connected to Earth by a conductor it is said to be **earthed**.

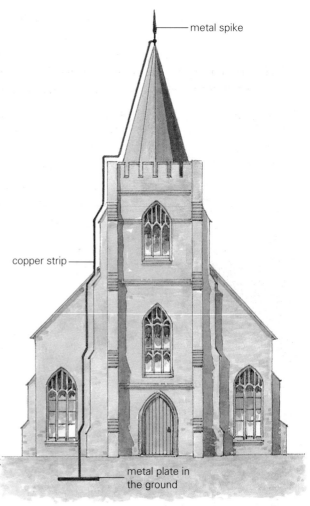

A lightning conductor allows the charge to flow to Earth without damage to the building.

Making use of static electricity

Photocopiers transfer an image onto paper by a method involving electrostatic attraction.

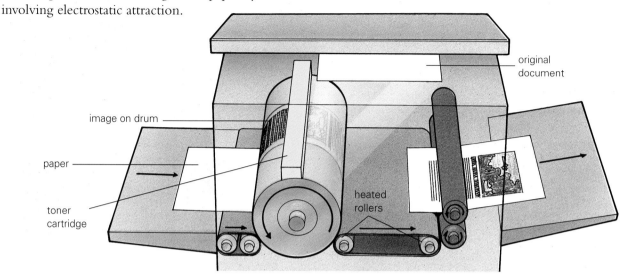

Inside a photocopier.

The positively charged surface of the drum is coated with a material called selenium, which becomes conducting when light falls on it. The parts that are lit will therefore lose their charge and so will not attract the negatively charged 'toner' powder. The toner is transferred from the drum to the positively charged paper, producing an image. The paper is then passed over heated rollers to melt the toner which sticks to the paper.

Removing soot by flue-ash precipitation.

Coal-burning power stations emit fumes containing soot particles which pollute the atmosphere. The soot can be removed by an electrostatic method called 'flue-ash precipitation'. The soot particles are negatively charged by passing over wires with large negative charges. They are then attracted to positively charged plates. The soot sticks to these plates, which can then be cleaned.

When metal railings are spray-painted, much of the paint can be wasted. The waste can be reduced by using an electrostatic method. The droplets of paint become charged by friction when they come out of the spray nozzle, and are attracted to the railings. As well as saving paint, this also speeds up the painting because the spraying no longer has to be done from all angles.

QUESTIONS

1. Explain how tall buildings are protected by lightning conductors.
2. Describe how the properties of static electricity are used in photocopiers.
3. Why do you think petrol pumps could be hazardous if they were made entirely from plastics? How does the use of metal make them safer?

1.3 Electric current, voltage and circuit diagrams

Current
When an electric current flows through a conductor, energy is transferred. One obvious effect is that the conductor often becomes warm. A current can have three effects:
- a chemical effect (see 1.8)
- a heating effect (see 1.9)
- a magnetic effect (see 3.1).

The symbol I is used to denote current. The size of a current is measured in **amperes** (symbol A), sometimes abbreviated to 'amps'.

Voltage
For an electric current to flow through a conductor there must be a voltage, really a potential difference (p.d.), across its ends. A voltage can be provided by a **cell**, a **battery** (made up of several cells), or a **generator** (or **dynamo**; see 3.4). This is the source of energy.

The symbol V is used to denote voltage or p.d. The size of a voltage or p.d. is measured in **volts** (symbol V).

'Voltage' and 'volts' are named after the eighteenth-century Italian scientist Alessandro Volta, who 'discovered' potential difference and was the first to produce a current from a chemical reaction.

Alessandro Volta, 1745–1827.

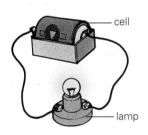
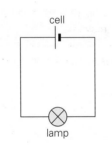

A complete circuit including an energy source is needed for a current to flow.

Circuits
For a current to flow the conductor must form part of a complete **circuit** – all conducting **components** must be connected by conducting wires, without a break, to the source.

When a current is flowing, energy is transferred from the source to the components. For example, in a simple circuit consisting of a cell and a lamp, energy which was stored in the cell is transferred to the lamp. The filament in the lamp is heated by the electric current to produce light.

The size of the current which flows in a particular circuit depends on the voltage provided by the cell. If the voltage is increased, then the current will increase. The voltage can be increased by connecting two or more cells end-to-end.

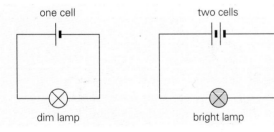

Increasing the voltage increases the current.

Circuit diagrams
Circuit diagrams like those above show you how the components that make up a circuit are connected together. They do not give a picture of the actual wiring – all connections are shown as straight lines with right-angled corners, and are drawn as short as possible, regardless of the actual length of wire used. Standard symbols are used for components.

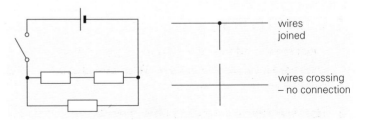

Note the convention used for showing that wires connect at junctions.

You will need to know the symbols for the following:

Switch	A switch operates by making and breaking an electrical circuit.
Cell	A cell stores chemical energy which is transformed into electrical energy when it is connected to a circuit.
Battery	A battery is made up of several cells.
Resistor	Resistors are used in circuits to reduce the current to a particular value, or to provide a particular voltage across part of the circuit.
Variable resistor	A variable resistor allows the value of the resistance to be changed to provide a range of values for current or voltage.
Lamp	A lamp converts electrical energy to light energy for illumination, or for signalling.
Fuse	A fuse is a component which is placed in a circuit for protection. It is designed to 'blow' and break the circuit when a specified current is exceeded.
Diode	A diode will only allow current to flow one way. It flows in the direction of the arrow on the symbol.
Voltmeter	A voltmeter measures the voltage between two points in a circuit.
Ammeter	An ammeter measures the current flowing through it.
Alternating current	An alternating current is an electric current that reverses its direction at regular intervals.

Resistors.

Variable resistors.

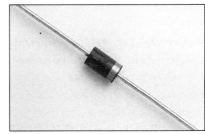

A diode.

QUESTIONS

1. What are the effects of an electric current? What is needed for a current to flow? What unit is used to measure electric current?

2. How is a voltage produced? How is electric current affected by changes in voltage? What unit is used to measure voltage?

3. Draw a circuit diagram for a circuit containing a cell, a variable resistor, a lamp and a switch.

1.4 Resistance and Ohm's Law

Resistance
All electrical conductors resist the current through them to some extent. This property is called **resistance** (symbol R) and it affects the size of the current flowing in a circuit. When the resistance is increased, the current will decrease. Resistance is measured in **ohms** (symbol Ω).

The resistance of a straight conductor (i.e. a wire) depends on:
- its length – the resistance increases with increasing length
- its thickness – the resistance is greater for thinner wires
- its material – good conductors have a lower resistance.

Components whose only purpose is to provide resistance are called **resistors**.

The size of the current which flows is determined by the voltage provided by the energy source, and by the resistance of the components and connections which make up the circuit.

Measuring current and voltage
The current flowing through part of a circuit can be measured using an **ammeter**. The ammeter must be connected in the circuit so that the same current flows through it as through the part of the circuit in question. This is called a **series** connection.

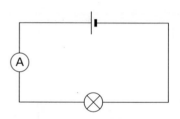

The ammeter is in series with the lamp.

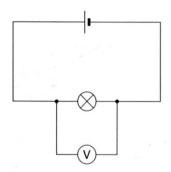

The voltmeter is in parallel with the lamp.

The voltage between two points in a circuit can be measured using a **voltmeter**. The voltmeter must be connected across, or between, the two points. This is called a **parallel** connection.

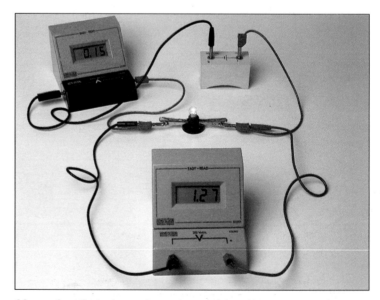

Measuring the current through and the voltage across a lamp.

Ohm's Law
The size of the electric current flowing through a conductor depends on the voltage across it, and the resistance of the conductor. When the voltage is increased, the current increases. For a particular metallic conductor, *the current is directly proportional to the voltage, provided the temperature remains constant.* This is **Ohm's Law**, formulated by the German physicist Georg Ohm in the 1820s. It means that if the voltage is doubled, the current will be doubled.

Georg Ohm, 1787–1854.

The current, the voltage and the resistance of a conductor are related by the following equation:

voltage (V) = current (A) × resistance (Ω)
or, in symbols: $V = I \times R$ or $V = IR$

This defines resistance. For **ohmic conductors** the resistance is constant and Ohm's Law of proportionality holds.

The relationship above can be expressed in two other ways:

$I = \dfrac{V}{R}$ and $R = \dfrac{V}{I}$

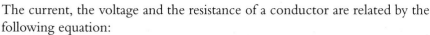

This diagram may help you to remember the formula for Ohm's Law. Cover up the quantity you want to find and you will see how to calculate it.

Examples

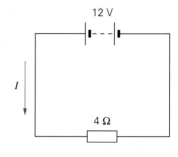

Current $I = \dfrac{V}{R}$
$= \dfrac{12\,V}{4\,\Omega}$
$= 3\,A$

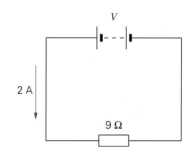

Voltage $V = I \times R$
$= 2\,A \times 9\,\Omega$
$= 18\,V$

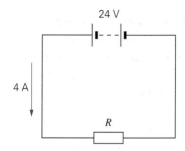

Resistance $R = \dfrac{V}{I}$
$= \dfrac{24\,V}{4\,A}$
$= 6\,\Omega$

QUESTIONS

1 What factors affect the size of the current flowing in a circuit?

2a What unit is used to measure resistance?
b What factors affect the resistance of a straight conductor?

3 Calculate the missing values in the circuits on the right.

4a Describe, with diagrams, how ammeters and voltmeters are connected into circuits to measure current and voltage.
b How would you use an ammeter and a voltmeter to find the resistance of a length of wire?

a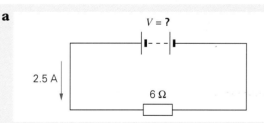

b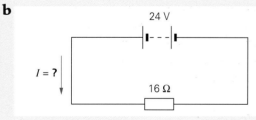

c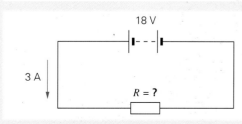

Circuits for question 3.

1.5 *I–V relationships*

According to Ohm's Law, if the temperature remains constant, the current through a metallic conductor is proportional to the voltage across its ends.

You can investigate this relationship for a length of resistance wire by using a variable voltage power supply, an ammeter and a voltmeter, connected as shown. For each voltage setting, measure the current through the wire and the voltage across it. If you plot a graph of current (I) against voltage (V), you should obtain a straight line, showing that they are proportional – the resistance wire is an ohmic conductor.

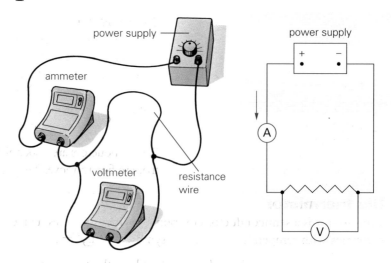

Circuit for investigating the current–voltage relationship for a length of resistance wire.

The slope of the line depends on the resistance of the wire – you can verify this by repeating the measurements for a different length of wire. Since $I = V/R$, the slope or gradient of the *I–V* graph is $1/R$. A longer piece of wire (with greater resistance) will therefore give a line with a smaller (less steep) slope. Similar length wires of different materials will give graphs of different slopes. A very good conductor (with small resistance) will give a line with a large (steep) slope.

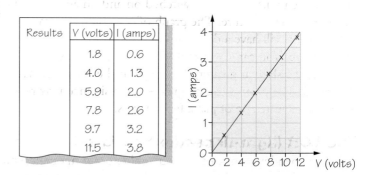

Results:

V (volts)	I (amps)
1.8	0.6
4.0	1.3
5.9	2.0
7.8	2.6
9.7	3.2
11.5	3.8

Typical results.

Resistance and temperature

The resistance of conductors changes with temperature. When a metallic conductor is heated, its resistance increases. This increase is slight, although it is greater for pure metals than it is for alloys. Resistance wire is usually made of an alloy, so that there is very little variation in its resistance, even over a wide range of temperatures. Some resistors, such as those used in electronic circuits, are made of carbon. When carbon is heated, its resistance *decreases*, but the change in resistance is very small.

Filament lamp

The temperature of any current-carrying wire will increase as the current transfers energy from the source (the power supply) to the wire. For a resistance wire the temperature change is small, and so its resistance is not affected. The filament of a lamp, however, can undergo a temperature change of several hundred degrees. This will produce a significant change in the resistance of the filament, and so the lamp is a **non-ohmic conductor**. If a graph of current against voltage is plotted, a curve will be produced. The resistance of the filament changes, so the slope of the graph changes.

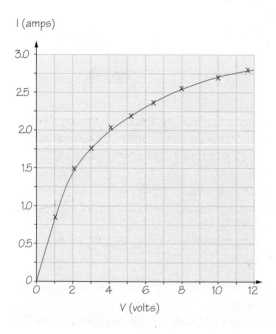

A filament lamp does not obey Ohm's Law.

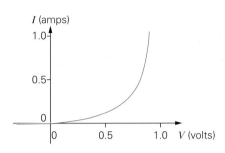

Current against voltage for a diode.

The diode

A diode is a **semiconductor** component used for its property of conducting electricity in only one direction. If a diode is reversed in a circuit it will not conduct. The graph of current against voltage for a diode shows that at low voltages the resistance is high, and then it decreases. At voltages greater than a particular value, the resistance becomes very low, so that a large current flows and the diode is damaged.

Semiconductors are a special class of materials which have few electrons that are free to move. They are non-ohmic even at constant temperature.

The thermistor

A thermistor is a semiconductor component which has a resistance that varies with temperature. Its circuit symbol is ─[/]─ .

The resistance of a thermistor *decreases* sharply as the temperature rises. Thermistors can be used as thermometers, and as part of electronic circuits which are switched on and off automatically by a change in temperature. The graph of current against voltage for a thermistor will have a different slope depending on the temperature. You can see that at higher temperatures (larger currents) the slope is greater, showing that the resistance has decreased. The change can be great – the resistance may be halved for a temperature rise of just a few degrees.

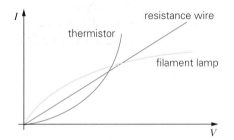

As the current increases the temperature rises, causing the resistance of the thermistor to decrease, while that of the filament lamp increases and that of the resistance wire remains constant.

The LDR (light-dependent resistor)

A light-dependent resistor is a semiconductor component which has a resistance that varies with the level of light. Its circuit symbol is ─⊖─ .

As the light intensity rises, the resistance of the LDR falls. The most common type has resistance which ranges from $10\,M\Omega$ in darkness to about $100\,\Omega$ in bright sunlight, and is about $5\,k\Omega$ in normal room lighting. LDRs can be used to measure light intensity, and as part of electronic circuits which are switched on and off automatically by a change in light intensity.

QUESTIONS

1 Use the data below, for an unknown conductor, to draw a graph of current I against voltage V. From your graph, do you think it is an ohmic conductor? Explain your answer.

Voltage V (V)	Current I (A)
0	0
2	0.4
4	0.8
6	1.2
8	1.6
10	2.0

2 The graph opposite of I against V for a filament lamp shows the results of an experiment with a filament lamp designed to take a current of 3 A from a 12 V supply.
  a Explain the shape of the graph.
  b Calculate the resistance of the filament when the lamp is working normally.
  c Estimate from the graph its resistance when the current is 1 A.

3a What is a thermistor? How does its resistance vary as it warms up?
  b What is a light-dependent resistor (LDR)? How does its resistance vary?

1.6 Series and parallel circuits

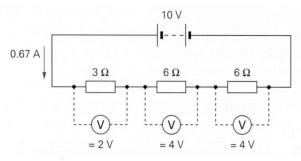

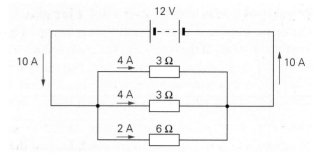

In a series circuit the supply voltage is 'shared' between the components in proportion to their resistance.

In a parallel circuit each component has the same voltage across it so the current through each one can be calculated separately.

When components are connected in series:
- the same current flows through each component
- the sum of the voltages across all the separate components is equal to the supply voltage.

When components are connected in parallel:
- each component has the same voltage across it
- the total current through the whole circuit is equal to the sum of the currents through the separate components.

Calculating total resistance

When you want to calculate the current in a circuit with more than one component or device you will need to calculate the total resistance.

If a resistance is added in series the total resistance will increase, because the current has to pass through extra resistance. The total resistance in a series circuit is the sum of all the separate resistances.

If a resistance is added in parallel, the total resistance will decrease, because the current now has an extra pathway available. For two equal resistances in parallel, the total resistance will be half the value of one resistance. In general, for resistances $R_1, R_2, R_3, ...$, in parallel the total resistance R_t can be found from:

$$\frac{1}{R_t} = \frac{1}{R_1} + \frac{1}{R_2} + \frac{1}{R_3} + ...$$

If there are *two* resistors in parallel, the total resistance is given by the formula:

$$R_t = \frac{R_1 \times R_2}{R_1 + R_2}$$

A useful way of checking that your result is reasonable is to ensure that your total resistance is less than the value of the smallest resistance, because each resistance added in parallel will reduce the total resistance.

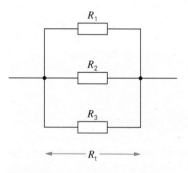

Resistors connected in parallel.

Connecting cells

If cells are connected together in series, the voltage provided is equal to the sum of their individual voltages.

If identical cells are connected together in parallel, the voltage will be just the same as for one cell. A slightly larger current will flow through the circuit, however, because the total resistance of the cells themselves will be reduced – even cells have resistance!

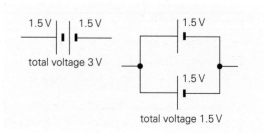

Practical series and parallel circuits

If lamps are connected together in series, the voltage of the energy source is 'shared' between the lamps. If the lamps are identical it will be shared equally and they will all have the same brightness. An increase in the number of lamps will, however, reduce the voltage across each one and so cause them all to become dimmer. The failure of one lamp will break the circuit and all the lamps will be switched off.

If the lamps are connected together in parallel, the voltage of the energy source is connected across each lamp. If they are identical they will all have the same brightness. Their brightness will not be affected if the number of lamps is increased. However, more lamps will mean that more current in total is taken from the energy source – a battery would be exhausted more rapidly. When lamps are connected in parallel the failure of one will not affect the others.

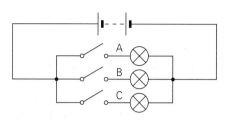

Lamps A, B and C can be switched on and off separately by the switches in parallel.

The lights in your house are connected to the supply in a parallel circuit, so that you can have some lights on and some off.

Car lighting also uses parallel connection of lamps, so that there is a fixed voltage (that of the battery) across each lamp. The lamps are designed so that they provide the required illumination at a particular voltage.

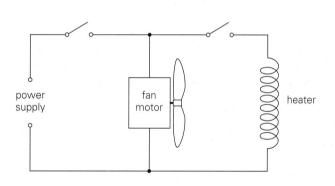

Sometimes it is desirable to have components in parallel – so that there is maximum voltage across each – but be able to switch them all on or off together. A switch is then connected in series with the components.

A fan heater has switches in series. For safety, the heater cannot be on unless the fan is on.

QUESTIONS

1. Draw two circuit diagrams, each with a 6V battery and two 3Ω lamps, to show the difference between a series and a parallel circuit.

2. A 12V battery is connected across two 6Ω resistors connected in series. Draw a circuit diagram, and calculate the current through the resistors.

3. Two 6Ω resistors are connected in parallel across a 12V battery. Draw a circuit diagram, and calculate the current through each resistor.

4. Three 6V batteries are connected in series with two identical resistors. A current of 2A flows in the circuit. What is the value of each resistor?

5. Calculate the current through the 3Ω resistor in this circuit.

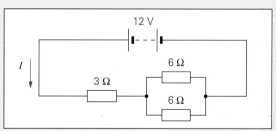

6. Draw a circuit using a battery, two lamps, and three switches so that the lamps can be switched on and off separately, and so that both lamps can be switched off together by one switch.

1.7 Movement of charge

When an electric current flows through a metallic conductor, the moving charges are the negatively charged 'free' electrons in the metal.

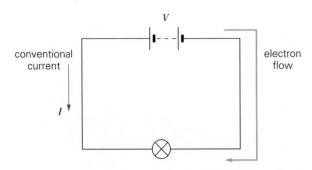

Electrons flow in the opposite direction to the conventional current.

The voltage needed to cause a current to flow can be provided by one or more cells. Chemical reactions inside a cell cause one terminal to become electrically positive with respect to the other terminal. When a circuit diagram shows the direction of an electric current, it is shown moving from the positive to the negative terminal of a cell or battery. This is the direction in which positive charges would move, and is known as 'conventional current' flow. In fact, electrons carry negative charges from the negative to the positive terminal – in the opposite direction to the arrows shown on circuit diagrams. The reason for this discrepancy is historical – Benjamin Franklin got it wrong!

Charge is measured in units called **coulombs** (C). One coulomb is equivalent to the charge on 6.25×10^{18} electrons. The unit is named after the French physicist Charles Augustin de Coulomb, who investigated and measured accurately the very small force between electric charges.

The flow of charge when a voltage is applied is continuous – in the same way that water pumped through a network of pipes cannot build up at any point, the charge in a circuit cannot build up.

If you measure the current in a simple series circuit with an ammeter, it will have the same value wherever you connect the ammeter in the circuit. If you measure the current at various points in a circuit with parallel connections, you will find that the current flowing into any connection is equal to the current flowing out of the connection.

Charles Augustin de Coulomb, 1736–1806.

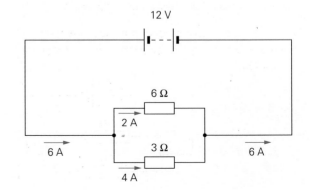

Current is the *rate at which charge flows* in a conductor. It can be calculated by using the following relationship:

current (A) = $\dfrac{\text{charge (C)}}{\text{time (s)}}$

or, in symbols: $I = \dfrac{Q}{t}$

where Q stands for charge and t for time.
This relationship can also be expressed as:

$Q = I \times t$ or $t = \dfrac{Q}{I}$

So, when a current of one ampere is flowing, one coulomb of charge moves past a given point in one second. A larger current means that more charge passes a given point per second.

Voltage, charge and energy

The energy source for an electrical circuit provides a voltage which causes the charges to move. The size of the voltage determines the rate of charge flow, i.e. the size of the current in a particular circuit, assuming that the resistance of the circuit remains constant. The voltage of the source provides the **energy** needed to move the electrons around the circuit. Energy is measured in **joules** (J). Voltage can be defined as the *energy transferred per coulomb of charge*:

voltage (V) = $\dfrac{\text{energy (J)}}{\text{charge (C)}}$

or, in symbols: $V = \dfrac{E}{Q}$

A cell which has a voltage of one volt will transfer one joule of energy to each coulomb of charge. A higher voltage will give more energy to each charged particle, so the particles will move faster and larger numbers will pass a given point every second. This is why increasing the voltage increases the current in a circuit.

The voltage across a circuit component is the energy transferred by a coulomb of charge *to* the component.

QUESTIONS

1 Describe how current flows in a metallic conductor. What is meant by 'conventional current'?

2a Calculate the current flowing through each resistor.
b How much charge passes through each resistor in 1 minute?

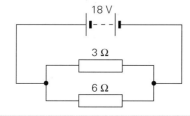

3 What is the relationship between charge and current? If a current of 2A flows in a resistor, how much charge passes through in one minute?

4 What is the relationship between voltage, energy and charge? If a battery gives 24 J of energy to 4 C of charge, what is the voltage of the battery?

5 If a battery connected across a circuit is replaced by another with a higher voltage, how will the movement of the electrons in the circuit be affected?

1.8 Electrolysis; cathode ray oscilloscope

Ions and electrolysis

An electric current can be conducted by some chemical compounds, called **ionic** compounds, when they are melted or dissolved in water. The compounds break up into positive and negative charged particles called **ions**. When a voltage is applied the ions are free to move. The current is the flow of these ions. The voltage is applied to a molten substance or solution, called the **electrolyte**, through terminals called **electrodes**. The process is called **electrolysis**. The positive ions move to the negative electrode, or **cathode**, and the negative ions move the positive electrode, or **anode**. New substances are deposited or, if gaseous, given off at the electrodes – the electric current has a **chemical effect**. For example, if the electrolyte is a metal compound, the metal is deposited at the cathode. This process is therefore used for extracting metals from their ores, purifying metals, and plating one metal with a thin layer of another metal.

During electrolysis the mass of the substances deposited or liberated at the electrodes is proportional to the charge which has passed, so it will depend on the current and the time.

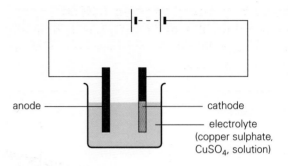

When an electric current is passed through copper sulphate solution, pure copper is deposited at the cathode. This process can be used for copper-plating a metal.

Aluminium is extracted from its ore, bauxite, by electrolysis of the molten ore. Pure aluminium is deposited at the cathode. This process uses a lot of energy, so it is important that we recycle aluminium whenever possible.

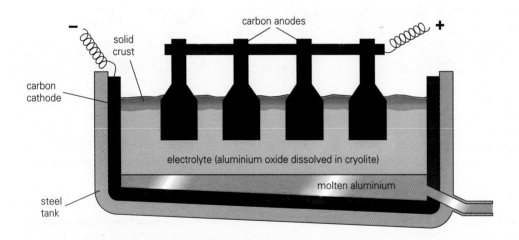

Aluminium is extracted from its ore by electrolysis. Its ore, bauxite, is purified, to provide aluminium oxide, which is then dissolved in molten cryolite, another ore of aluminium, to form the electrolyte. The electrolyte is kept at a temperature of about 1000°C. Pure molten aluminium is produced at the cathode. This process uses a lot of electrical energy, so it is important that we recycle aluminium whenever possible.

Cathode ray oscilloscope (CRO)

A **cathode ray oscilloscope (CRO)** is used for measuring and displaying voltages and currents (see pages 20 and 35). The movement of charge – electrons – is controlled by a number of applied voltages.

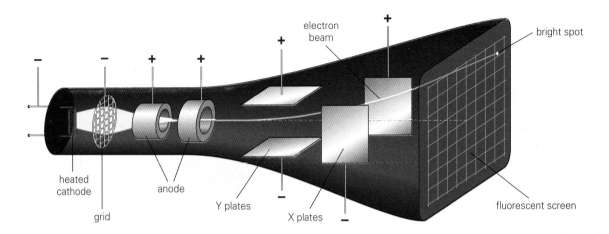

An **electron gun** produces a beam of electrons in an evacuated tube. This beam produces a bright spot on the screen which is coated with a fluorescent material. The electron gun consists of a cathode, a grid and an anode. The cathode is a heated filament which emits electrons. The electrons are accelerated by the positive voltage on the anode, which also focuses the beam. The grid affects the number of electrons that reach the screen, and is used to control the brightness. The X and Y plates control the deflection of the spot from the centre of the screen in the horizontal and vertical directions. Voltages to be measured are applied to the Y plates – the size of the deflection is proportional to the voltage applied. A **time base** is applied to the X plates. This moves the spot across the screen and then flicks it back to start again, so that voltages which vary over time can be displayed.

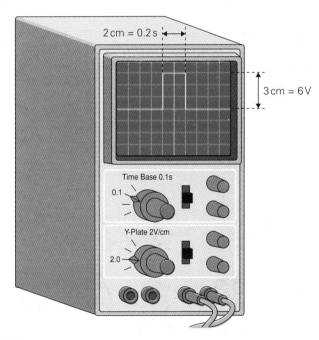

A CRO.

QUESTIONS

1. Explain briefly the following terms: ions, electrolyte, anode, cathode.

2. When electrolysis is used to plate one metal with another, what factors affect the quantity of metal deposited? To obtain an even coating of metal, the plating should be slow. How would you ensure this?

3. Why are aluminium smelters often located near hydroelectric power stations or nuclear power stations?

4. Why does the extraction of aluminium use so much energy?

5. What is the purpose of a time base in a CRO?

1.9 Electrical power

Electrical **power** (P) is the *rate at which energy is transferred* in an electrical device. It is measured in **watts** (W). One watt is equivalent to one joule per second. Power can be calculated from the relationship:

power (W) = $\dfrac{\text{energy transferred (J)}}{\text{time (s)}}$

or, in symbols: $P = \dfrac{E}{t}$

The rate of energy transfer, or power, in an electrical device depends on the current through it and the voltage across it. The relationship is:

power (W) = current (A) × voltage (V)

or, in symbols: $P = I \times V$ or $P = IV$

Consider the heating element shown. The power, or the rate at which the electrical energy is converted to heat in the element, can be calculated:

$P = I \times V$
$= 2\,A \times 12\,V$
$= 24\,W$

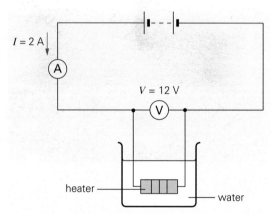

This means that 24 joules of electrical energy are converted in the element in every second.

The total amount of energy transferred can also be calculated, if the time for which current is flowing is taken into account. Combining the relationships

power (W) = current (A) × voltage (V)

and power (W) = $\dfrac{\text{energy (J)}}{\text{time (s)}}$

we obtain

energy (J) = current (A) × voltage (V) × time (s)

If the current in the circuit above is flowing for 5 minutes, the energy consumed in the element to heat the water can be calculated:

energy = 2 A × 12 V × 300 s
= 7200 J

It is important to use the correct units in such calculations. The time must be converted to seconds.

For large amounts of energy and power, the units **kilojoules** (kJ) and **kilowatts** (kW) are often used. One kilojoule is equivalent to 1000 joules, and one kilowatt is equivalent to 1000 watts or 1000 joules per second.

Electrical heating

When electrical energy is transferred in a conductor, it may be converted to various forms of energy, depending on the device or appliance. Some of the energy, however, always ends up heating the conductor. This may be why the electricity is being used, as in a heater, or a hairdryer, or the heat may be a by-product of the use of the electricity, as in an electric drill or a loudspeaker. This **heating effect** is due to the resistance of the material of the conductor to the flow of electrons. In a metal wire and in a resistor, *all* of the electrical energy transferred is converted into heat.

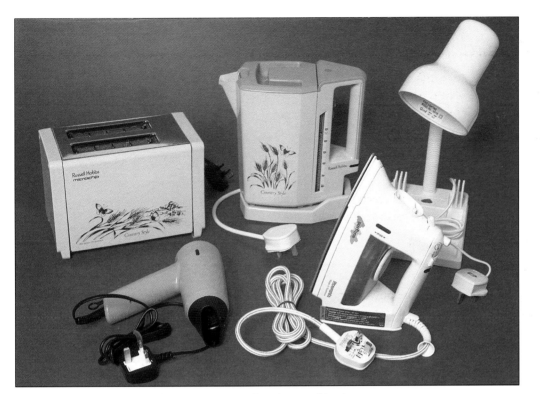

These appliances convert electrical energy into heat – they have a high resistance.

The rate at which energy is transferred in any component, or its power, can be calculated from:

$P = I \times V$

For a resistance R this can be expressed in another way by combining it with the Ohm's Law relationship:

$V = I \times R$

Substitution gives:

$P = I \times I \times R$
$ = I^2 \times R$

This shows that the rate at which heat is produced in a resistance wire or a resistor depends on the square of the current, and on the value of the resistance. If the current is doubled, the heat produced will be four times greater. A larger resistance will produce more heat than a smaller one carrying the same current.

This relationship is important when materials are being chosen for electric circuits. The connections between components should be made of materials with low electrical resistance. This ensures that as little energy as possible is wasted in heating the connections between the components.

QUESTIONS

1a What units are used to measure **i** energy, **ii** power? What is the relationship between energy and power?

b How would you calculate the power of an electrical device?

2 A car has a 12V battery, and its headlamps which are in parallel have a power rating of 60W.

a What is the current flowing through the filament of the headlamp?

b How much energy will be used by a headlamp in 10 minutes?

c What is the resistance of the filament of the headlamp?

3 A personal stereo has a power rating of 1W. It takes two 1.5V cells which each store 10 000J of energy. How long will it play before cells should be replaced?

4 A 48W water heater takes a current of 4A.

a What voltage supply should it be connected to?

b What is the resistance of the element?

c How much energy will it supply to heat water if it is left on for 20 minutes?

CHAPTER 2: ELECTRICITY IN ACTION

2.1 *Mains supply*

Your home will probably have dozens of appliances that use electricity as their source of energy; these may range from cookers, washing machines and lights, to televisions, video recorders and computers. To make these work you plug them into the 'mains' supply. You may use things such a personal stereo, a calculator, computer games or a camera which use batteries as their energy source. When a battery is used the current always flows in the same direction – from the positive to the negative terminal of the battery. This is called **direct current** (or **d.c.**). The mains electricity supply in the UK is **alternating current** (or **a.c.**). This means that the direction of the current is continually reversing.

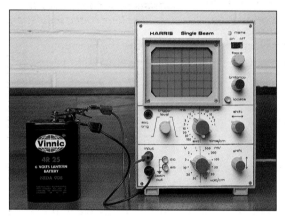

A 6 V d.c. input gives a straight line above or below the centre of the screen.

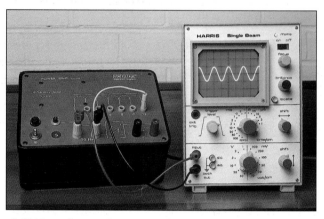

A 12 V a.c. input gives a regular waveform, symmetrical above and below the central (zero) line on the screen.

You can see the difference between a.c. and d.c. using a cathode ray oscilloscope. The display shows how the voltage changes with time. A battery connected to an oscilloscope gives a trace which is a straight line. An a.c. supply (*not* mains, as the voltage is too large to experiment with, but from the a.c. terminals of a laboratory power supply) shows a voltage that is continually changing from positive to negative. The current follows a similar variation.

Mains voltage in the UK is quoted as 230 V. If you could look at how this voltage varied with time, you would see that it changes from 0, to 325 V, to 0, to 325 V in the opposite direction, and back to 0 over one-fiftieth of a second. The heating effect of this varying voltage is the same as that of a direct voltage of 230 V, so this is why this figure is used. It provides a useful 'average' value for the voltage of the a.c. supply. A light bulb connected to a d.c. supply of 230 V (roughly 20 car batteries connected in series) would be lit to the same brightness as it would when connected to the mains a.c. supply. The **frequency** of the mains supply is 50 Hz – this means that there are 50 complete to-and-fro 'cycles' every second, so the direction of the current reverses 100 times every second.

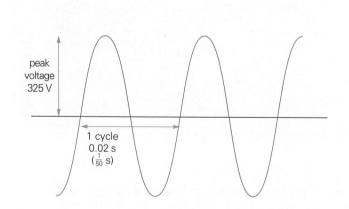

The waveform of the mains voltage.

Be safe!

When you make a connection to the mains supply, the circuit is made by the **live wire** and the **neutral wire**. When the live wire is at a positive voltage, it takes the current to the appliance; the neutral wire takes the current from the appliance back to the supply. When the voltage is reversed, the current flows the other way. The neutral wire remains at 0 V, and the voltage of the live wire varies from +325 V to -325 V. This high voltage could be fatal, so the live and neutral wires are always insulated by outer layers of plastic.

The fixed wiring in a house is made of thick copper wires inside plastic insulation – red for the live wire and black for the neutral wire. These cables can be bent to shape, but do not have to be very flexible because they stay in the same position.

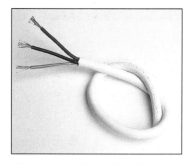
Three-core flex is flexible – it consists of stranded copper wire in colour-coded insulation.

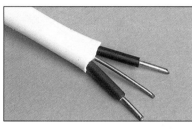
Mains cable consists of thick copper wires. The wire with no insulation is the 'earth' connection; when wired up the earth is always insulated with green/yellow sleeving.

The 'flexes' which connect electrical appliances to the mains supply through plugs need to be very flexible. They consist of many thin strands of copper wire, insulated with soft plastic.

Any switch in a mains circuit, whether it is on the socket or in the appliance, must be in the live wire. This is so that the voltage is zero in all wires beyond the switch when the appliance is switched off.

Mains wiring can carry large currents, and so will be heated. Flexes and cables should always be thick enough to carry the current without overheating.

The following precautions should be taken to avoid danger when using the mains supply.

- Never overload a socket with too many appliances, or the wiring could overheat and start a fire.
- Ensure plugs and flexes are in good condition. There should be no breaks in the insulation which may expose bare metal. Twisted or frayed cables may lead to damaged insulation.
- Ensure the fuse in a plug (see 2.2) is of the value recommended by the manufacturer.
- Ensure all connections are firm. Flexes which are too long can easily be pulled or tripped over.
- Never poke metal objects into sockets.
- Keep sockets dry – water can sometimes conduct electricity, and wet skin can make the human body a good conductor!

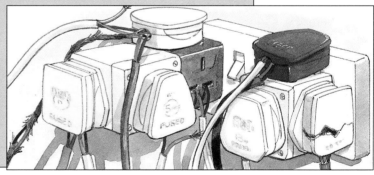
How many hazards can you spot?

QUESTIONS

1. Sketch a graph to show how an a.c. voltage varies over a complete cycle. What is the frequency of a.c. in the UK?

2a. What are the names given to the two wires which provide the mains supply?

b. Which wire is used for switching? Why is this important?

3. What are the differences between the cable used for the fixed wiring in a house, and the flex which connects an appliance to the supply?

2.2 Household wiring

Into our homes

The mains supply to a house comes through a **service cable**. This runs through the **main fuse** to the **electricity meter**, which measures the amount of energy used, and then to the **consumer unit**. The supply is earthed by connection to a metal spike which is embedded in the Earth. The Earth is at 0 V, so the **earth wire** provides an alternative to the neutral wire (in case of a fault) for the return current.

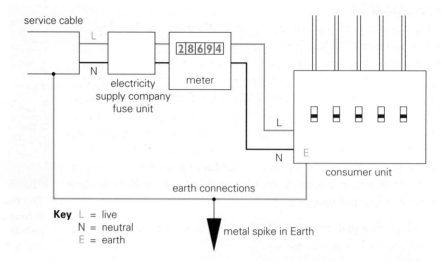

How electricity is supplied to a house.

The consumer unit has either **fuses** or **circuit breakers**, each of which protects a particular circuit in the house from overload. They break the circuit if the current becomes dangerously large. Some appliances which use a large current, like cookers or water heaters, have their own fuse or circuit breaker in the consumer unit.

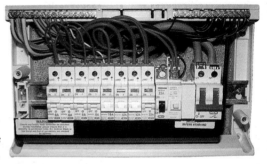

Domestic consumer unit. The circuit breakers (which look like switches) can be re-set easily.

Wiring sockets and lights

The power sockets in a house must be connected in parallel with each other, so that they all provide the same voltage and so that one can be in use while others are off. They are not all wired individually – they are connected to a **ring main**. This is a loop which consists of live, neutral and earth cables, to which 'spurs' are attached leading to each socket. Each ring main is fused at the consumer unit. A house will have two or three rings.

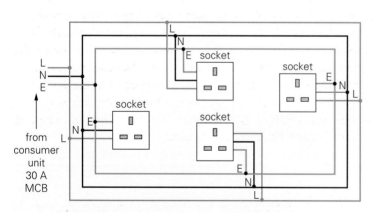

A ring main (MCB = miniature circuit breaker).

Lighting circuits are not rings – a live and a neutral cable (no earth) run to several 'junction boxes' around the house. The lamps and switches are connected to the junction boxes in parallel.

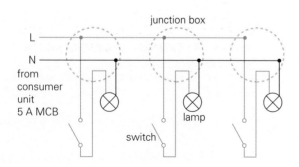

A lighting circuit.

Plugging in

All appliances with a power rating of 3 kW or less can be plugged into a household socket. The plug is wired as shown in the photograph. The fuse in a plug should be of the value recommended by the manufacturer. Values are given in terms of a current size in amperes, usually 3 A or 13 A. If the current through the appliance becomes greater than the fuse value, the wire in the fuse will melt and the circuit will be broken. The correct fuse value should therefore be a current value a bit larger than the normal operating current of the appliance. The fuse protects the wiring in the event of a fault which allows too much current to flow, leading to overheating and danger of fire.

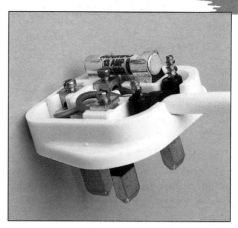

The live lead is brown, the neutral lead is blue and the earth lead is green and yellow. The fuse is connected in the live lead.

The plug's flex grip prevents a pull on the flex from disconnecting any of the leads from the pins.

There is always the risk when using a lawnmower of cutting through the flex. An RCD will protect you from shock.

Protection from shocks

If an appliance has a metal casing, the earth wire is connected to the casing. If a fault develops which causes the live wire to touch the casing, a large current will flow to Earth through the casing and so the fuse in the plug will 'blow' to cut off the supply.

Some appliances which are designed to be held by hand, such as a hairdryer, are **double insulated** by all parts being totally encased in plastic. No earth wire is then necessary.

Residual current devices (RCDs) can be used for further protection. They work by comparing the currents in the live and neutral wires. These should be the same, but if they differ by more than 30 mA (due to a fault, for example), the device will switch the current off within a few hundredths of a second. This will protect the user against electric shock. Your school science laboratory may well be equipped with RCDs. It is particularly important to use RCDs if you are working outdoors with an electrical appliance such as a lawnmower. You will have good contact with the Earth, and so are at more risk of current flowing through you.

QUESTIONS

1. What is the purpose of a fuse?

2. How does the earth connection in an appliance protect against electric shocks?

3a. Name the pins labelled A, B, C on this diagram of a plug. What is the colour of the lead attached to each pin?

   b. Explain the purpose of the part labelled D.

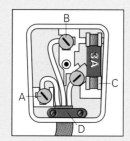

4. Explain why it would be dangerous if a plug was incorrectly wired so that the live and neutral leads were reversed.

5. Explain why you should use a residual current device (RCD) when clipping a hedge with an electric hedge-trimmer.

2.3 Domestic appliances

You use all sorts of electrical appliances to do different jobs in your home. They all convert electrical energy into other forms of energy to do their job. Most electrical appliances have 'rating' information which gives information on the voltage supply required, and the power consumption of the appliance. The information may be on a metal plate or in the case of a light bulb it is printed on the glass.

A filament lamp is the most common household electrical device. The filament is a long, thin wire made of tungsten, which is coiled, and then re-coiled. It has a very high resistance. When an electric current is passed through the filament, heat is produced and the filament's temperature rises to about 2500°C. This causes it to glow white-hot, and so produce light – although most of the electrical energy is transferred as heat.

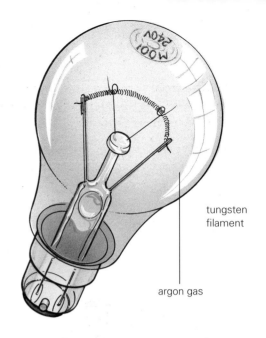

The electrical energy supplied to a refrigerator is converted to mechanical energy in the motor which pumps the refrigerant around the system, and pumps heat away from the interior to be dissipated by the cooling fins.

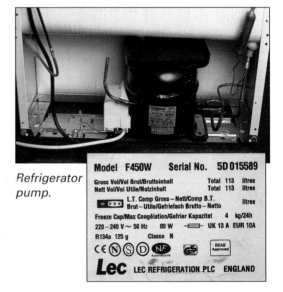

Refrigerator pump.

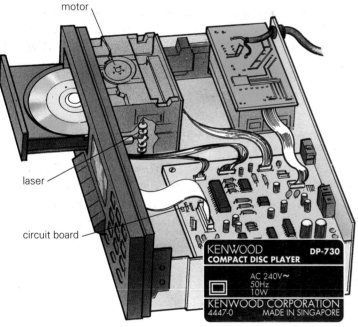

Electrical energy is transferred in many ways in a CD player – in the motor which turns the disc, in the laser which 'reads' the signals on the disc, and in the electronic circuits which amplify the signals so that they can be converted to sound.

Heating elements

The sole plate of an iron is heated by a heating element which is embedded in it. The current passing through the element heats it up – electrical energy is converted to heat. The earth wire is connected to the metal case.

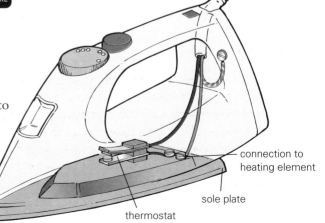

Heating elements are made from a high resistance wire, such as nichrome. They are designed to convert all of the electrical energy to heat.

A hairdryer converts some of the electrical energy into heat in the heating element, and some into mechanical (kinetic) energy in the motor. The conducting parts are double insulated. This is shown on the rating plate by ▢.

400W
50/60 Hz AP
MADE IN FRANCE

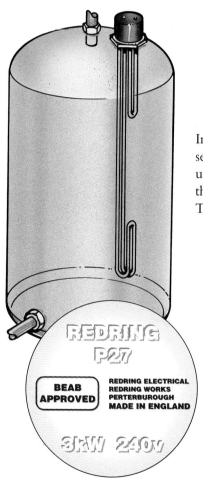

REDRING
P27

BEAB APPROVED

REDRING ELECTRICAL
REDRING WORKS
PERTERBUROUGH
MADE IN ENGLAND

3kW 240V

Immersion heaters use a heating element to heat the water in the tank. A second, shorter, element, which only heats the top part of the tank, can be used to give a rapid 'boost' to the temperature of the water in that part of the tank. The power rating of domestic immersion heaters is usually 3 kW. They are connected directly to the consumer unit.

Choosing a correct fuse

If you have to change the fuse in a plug, you should look at the rating plate on the appliance to find the power consumption. This will be measured in watts (W) or kilowatts (kW). If the power is up to 700 W, you should use a 3 A fuse; if the power is over 700 W (up to maximum of 3 kW), you should use a 13 A fuse. The value of the fuse is chosen so that it will carry the normal operating current, but it will 'blow' if the current becomes too high.

You can calculate the current taken by an appliance from its power rating:

$P = I \times V$
So $I = \dfrac{P}{V}$

Remember $V = 230\,\text{V}$ for the mains supply.

QUESTIONS

1 Using the information on the rating plates shown on these pages, calculate the current used by each of the electrical appliances, and suggest the most suitable value for the fuse in its plug. (Notice that older appliances may give the mains voltage as 240 V. You should use the value 230 V in your calculations.)

2 Why do some immersion heaters have double elements? Why is the shorter element placed at the top of the tank?

3 Why is tungsten used for the filament of a lamp? What difference would it make to the operation of the lamp if the filament was thicker?

2.4 The cost of electricity

Electrical energy is the energy of the charge which moves when a current flows. The power of an electrical appliance is the rate at which electrical energy is transferred. The electricity supply companies sell energy, but the joule is far too small a unit to be useful. On a bill from an electricity company, the 'units' of electricity used are **kilowatt-hours** (kWh). One kilowatt-hour is the amount of energy supplied in one hour to an appliance which has a power rating of 1 kW. It is equivalent to 3.6×10^6 joules (3 600 000 J).

WestSide ELECTRICITY
Faraday House
Amber Street
WestSide XZ9 7FG

Mrs J Courtney
12 Carfax Avenue, Martinford, Fryth QS3 3TT

Electricity Bill

Meter Readings		Units Used	Unit Price (pence)	VAT code	Amount £
Present	Previous				
01294	00035	1259	6.920	1	87.12

Standing Charges 1/6/96 to 31/8/96   10.95

Total Charges (excluding VAT)   98.07

VAT 1 £98.07 @ 8.0% Domestic   7.84

Total   105.91

E = Estimted reading   C = Your own reading

Account Number
042.3885/056.526

Date (Tax Point)
1 September 1996

Reading Date
1 September 1996

Business Use
0%

£105.91

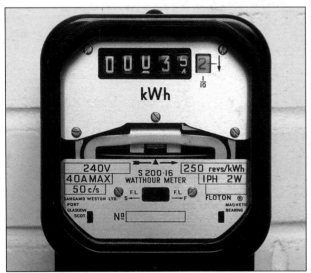
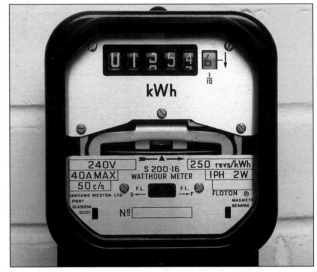

An electricity bill is calculated from the readings taken from the meter every three months. The cost of the electricity used is added to a fixed 'standing' charge.

Calculating the cost of an appliance

To calculate the cost of running a particular appliance, multiply its power in kW by the time used in hours by the cost per unit.

cost (p) = power (kWh) × time (h) × cost per unit (p)

Be careful with the units.

Typical power ratings of some appliances are shown below.

Table lamp	60 W	Fridge-freezer	150 W
Iron	1.2 kW	Microwave oven	1.5 kW
Kettle	2.2 kW	Food processor	500 W
Hairdryer	1000 W	Television	130 W
Hi-fi	200 W	Immersion heater	3.5 kW

Example: How much would it cost per month (30 days) to have an outside light (power 100 W) on a timer switch which was on for 10 hours per day? Take the cost of one unit as 7p.

cost = power × time × cost per unit

= 0.1 kW × (10 × 30) h × 7p

= 210p = £2.10

Off-peak electricity

The demand for electrical energy varies throughout the day. However, electrical energy cannot be stored and the production of electrical energy by power stations cannot be turned on and off very easily. The electricity supply industry tries to persuade customers to use **off-peak** electricity by making it cheaper. This is electricity supplied during part of the night, which is about half the price of daytime electricity. A special meter is installed in houses to measure the off-peak energy used. This can be an economical way to heat homes and hot water, if storage heaters and well-insulated water tanks are used.

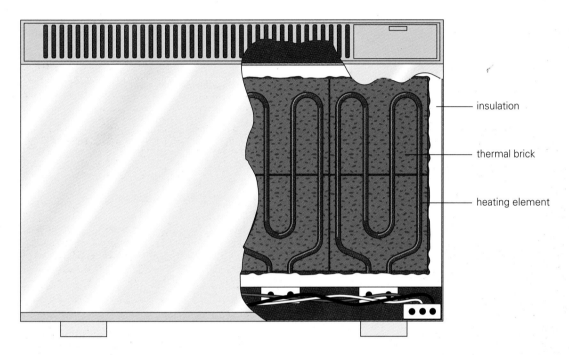

Thermal bricks inside a storage heater are heated by off-peak electricity in the night; they then warm the room in the daytime. The heat output can be controlled by opening and closing vents.

QUESTIONS

1. What is measured by an electricity meter? What units are used, and why?

2. Why do electricity companies sell off-peak electricity at a reduced price, although gas is the same price at all times of the day?

3. For each appliance in the table opposite, calculate
   a the cost of running it for one hour
   b how long you could run it for a cost of £1, taking the cost of one unit as 7p.

4. Calculate the cost of the following, assuming 7p per unit.
   a Two 100W light bulbs used for 5 hours per day for 90 days.
   b A 150W television used for 6 hours per day for 90 days.
   c A 1 kW iron used for 45 minutes twice a week for 13 weeks.
   d A 2.4 kW kettle used for 4 minutes, 15 times daily for 90 days.

5. A 10W radio takes 4 cells, total cost £1.50, and is left on continuously. It can also be operated from the mains. If the cells each store 10 000 J of energy, how long would they last before they should be replaced? How much would be saved if the mains supply was used for the same time, at a cost of 7p per unit?

CHAPTER 3: USING MAGNETS

3.1 Magnetism and electricity

Magnets

You probably know about some of the properties of magnets:

- Magnets are attracted to some metals – iron, steel, nickel and cobalt.

- Magnets are used as compass needles because they will always line up in a north–south direction if free to move.

- The ends of a magnet are called the **poles**, and they are described as the **north-seeking**, or just north (N), pole and the **south-seeking**, or just south (S), pole.

- The effects of a magnet are strongest at its poles.

- Magnets exert forces of attraction or repulsion on one another – like poles repel and unlike poles attract.

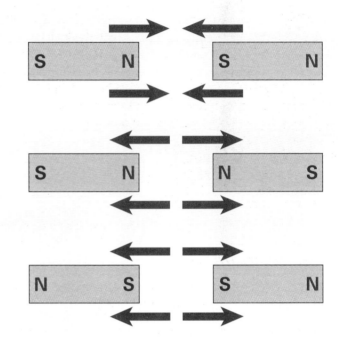

Like poles repel; unlike poles attract.

Magnetic fields

The space around a magnet in which a magnetic material experiences a force is called the **magnetic field**. It can be represented by imaginary arrowed lines, called **lines of force** or **field lines**, which show the direction of the force that would act on the north-seeking pole of another magnet. The force is strongest where the lines of force are closest together. A magnetic field can be detected with a compass, which will line up with the lines of force.

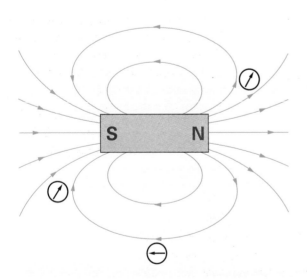

The magnetic field around a bar magnet in the plane of the paper – remember that the field is all around the magnet, above and below it as well as to the side.

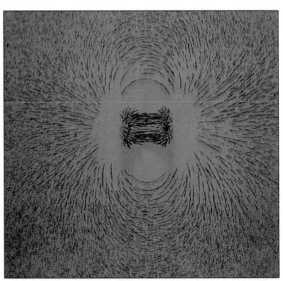

Sprinkling iron filings on to a sheet of paper with a bar magnet underneath can show the magnetic field pattern where the field is strong.

The magnetic effect of a current

In the 1820s André Marie Ampère, a French physicist, found that two close current-carrying wires exerted a force on one another. He was one of the first to discover the **magnetic effect** of an electric current – a magnetic field can be detected near a wire carrying a current. The unit of current is named after him because its value is defined in terms of the magnetic force it produces.

André Marie Ampère, 1775–1836.

The magnetic field around a current-carrying wire.

The lines of force around a straight current-carrying wire are concentric circles which run around the wire.

The direction of the field can be determined by the 'right-hand screw rule' – if a screw points in the direction of the (conventional) current, the direction of the field lines is shown by the way your hand must move to turn the screw.

Ampère also found that a current-carrying coil of wire behaved just like a magnet. If a wire is wound into a long cylindrical coil, called a **solenoid**, the magnetic field has a pattern very similar to that of a bar magnet.

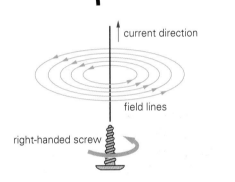

The right-hand screw rule.

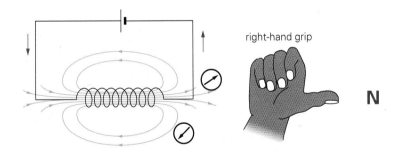

The magnetic field of a solenoid.

The 'right-hand grip rule' determines which end of the solenoid acts as a north pole – if you imagine your right hand gripping the solenoid with your fingers pointing in the direction of the current, your outstretched thumb points to the north pole.

The existence of a magnetic field around a current-carrying coil is the principle of an **electromagnet**.

QUESTIONS

1. Why are the poles of magnets referred to as N and S poles?

2. Sketch the magnetic field patterns for **a** a bar magnet, **b** a solenoid, and **c** a straight current-carrying wire. Show on your sketches the direction of the field (in relation to the poles/current direction), and make it clear where the field is strongest.

3. Explain why field lines never cross in a magnetic field diagram.

3.2 Electromagnets

An electromagnet consists of a coil of wire, through which a current can be passed, wrapped around a **soft iron core**. This core of magnetic material increases the strength of the field due to the coil. 'Soft' iron is easily magnetised, and easy to demagnetise – it does not retain its magnetism after the current is switched off. Steel, on the other hand, is hard to magnetise and demagnetise, and so it retains its magnetism. It is used for **permanent** magnets, sometimes with added aluminium, nickel or cobalt.

An electromagnet can be very strong, and has the added advantage over a permanent magnet that it can be turned off and on by switching the current off or on.

The strength of an electromagnet depends on:
- the size of the current flowing through the coil
- the number of turns on the coil
- the material in the core of the coil.

The poles of the electromagnet will be reversed if the direction of the current is reversed.

This scrapyard crane is picking up scrap iron and steel, which can be released when the power to the electromagnet is turned off.

Uses of electromagnets

Electromagnets are used in a wide variety of applications:
- for lifting scrap iron in scrapyards
- in the recycling of metals
- for removing metal splinters from eyes
- in the magnetic suspension systems of 'monorail' trains
- in many electrical devices such as electric bells, various switches, motors, loudspeakers, and telephone ear-pieces.

Germany's 'Magnetbahn' train is suspended above the track by a magnetic field between electromagnets on the train and electromagnets on the track.

The electric bell

An electric bell uses an electromagnet to attract an iron armature when the current is turned on. This strikes the clapper on the bell, and also breaks the circuit at the 'make and break' contact. The electromagnet is switched off, so that the armature springs back to its original position. The current flows again and the process is rapidly repeated to produce a continuous ringing.

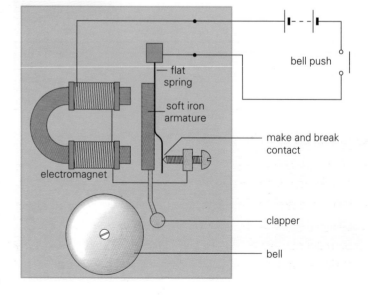

Electric bell.

The relay

A relay uses an electromagnet in one circuit to operate a switch in another circuit. A small current can then be used to switch a much larger current on and off. When a current flows through the electromagnet, it attracts an iron armature which closes the contacts in the second circuit. When the current through the electromagnet is switched off, the lever falls back and the contacts open again, breaking the second circuit.

Relays have a wide variety of applications – in car starter motors and headlights, in robotic and other heavy machinery, and in electronic devices.

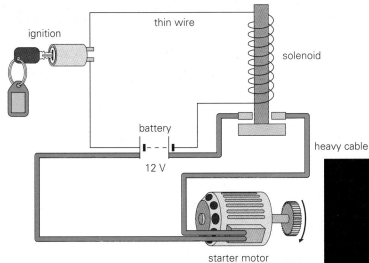

A car's starter motor takes over 100 A. It is operated by a relay when the ignition switch is turned.

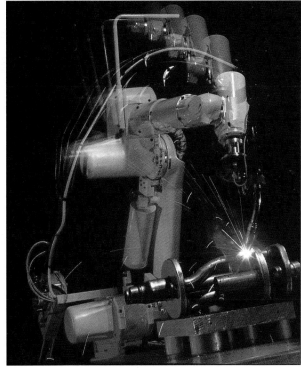

Relays are used in robotic machinery.

QUESTIONS

1. What are the factors which affect the strength of an electromagnet? How could you reverse the polarity of an electromagnet?

2. Which materials are suitable for making **a** a permanent magnet, and **b** the core of an electromagnet?

3. Powerful electromagnets are used on cranes in scrapyards to sort iron and steel items from other materials, and to move them around. Explain why a permanent magnet would not be suitable for this.

4. Why is a relay used to switch on a car's starter motor?

3.3 Electric motors

Because there is a magnetic field associated with an electric current, and a magnetic field can exert forces, it is possible to use this **electromagnetism** to move things. The **electric motor** is a vital part of many of the products of technology, from washing machines to satellites.

A wire carrying a current has a magnetic field around it, so if it is in the magnetic field of another magnet, there will be a force exerted on it. The size of the force depends on the size of the current, and the strength of the magnetic field. The direction of the force on the wire will be reversed if either the direction of the current or the direction of the magnetic field is reversed.

The direction of the force can be predicted using 'Fleming's left-hand rule'. If you hold the fingers of your left hand as shown, and your first finger shows the direction of the magnetic field while your second finger shows the direction of the current, then your thumb shows the direction of the motion produced by the force. The three directions are at right angles to each other.

The electric current needed to power the motors in this communications satellite is produced by panels of solar cells.

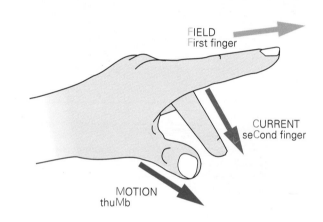

Fleming's left-hand rule.

Construction of a motor

A simple electric motor, such as one used in a toy car, consists of a flat coil of wire which is mounted between the poles of a magnet. When a current passes through the coil, the force exerted on it causes it to turn. The current from the (d.c.) power supply is fed into the coil through conducting **brushes** which make contact with the **commutator**. The commutator is a metal cylinder split into two halves, so that the direction of the current through the coil is reversed every half revolution. This change of direction of current causes the coil to continue to turn in the same direction.

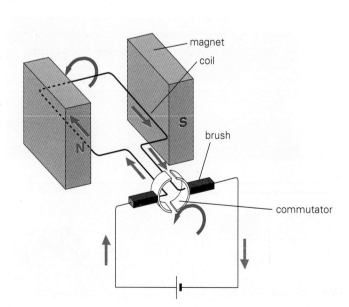

A simple electric motor.

In this simple motor the size of the force varies as the coil rotates – it is zero when the magnetic field of the coil is parallel to the external magnetic field. This causes a jerky rotation, and makes it easy to stall the motor. Motors which are used in industrial machinery or domestic appliances have several coils of wire arranged at different angles, and a multi-segment commutator, so that there is always one coil in a position to receive the maximum turning force. It is this coil which is in contact with the brushes and so has the current passing through it. This provides more power, and smoother running. The magnetic field is provided by electromagnets, which can produce a very strong field. The turning force on a coil depends on:
- the strength of the magnetic field
- the size of the current
- the number of turns in the coil.

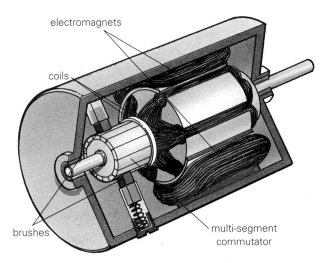

This powerful motor is from a washing machine. It has several coils.

Loudspeakers

Loudspeakers also work on the principle of the movement of a coil carrying a current in a magnetic field. The coil is attached to the cone of the loudspeaker and is mounted between the poles of a magnet. The magnet is specially shaped so that there is a very strong field in the narrow gap between its poles. The electrical signal, which consists of a rapidly varying current, passes through the coil which is in the magnetic field. The coil and cone therefore have a varying force exerted on them which causes them to move backwards and forwards very rapidly. The movement of the cone causes the air in contact with it to move, and this produces the sound. The vibrations follow the variations in the electrical signal, so that the sound produced matches the original sound of the speech or music.

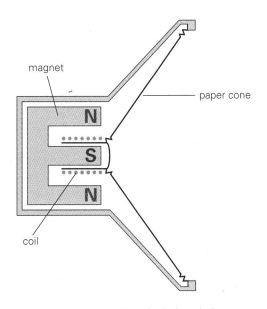

A loudspeaker converts electrical signals into sound.

QUESTIONS

1. Using Fleming's left-hand rule, say whether the wire in these diagrams would move towards you or away from you.

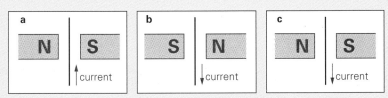

2. What is the purpose of the commutators in a motor?

3. How can electric motors be made more powerful?

3.4 Electromagnetic induction

In 1831 Michael Faraday found that when he moved a magnet in or out of a coil of wire an electric current was produced. You can verify this by using a coil connected to an ammeter that is able to detect small currents. You will see that a current flows when you move a magnet into the coil. This means a voltage is being produced — we say a voltage is **induced**, and the effect is called **electromagnetic induction**.

The size of the induced voltage depends on:
- the strength of the magnet
- the number of turns in the coil
- the speed of the movement.

The direction of the voltage (and hence of the induced current) depends on the direction of movement of the magnet, and which way round the magnet is.

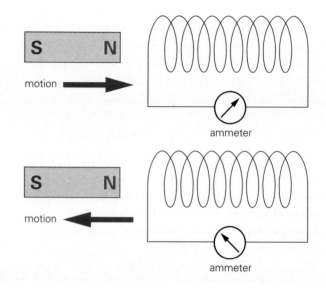

When you remove the magnet from the coil, the current flows in the opposite direction. What happens if you insert the S pole?

A voltage is induced whether it is the magnet or the coil which moves. It is the *relative motion between a conductor and a magnet field* that produces the effect — a changing magnetic field through a coil, such as that of an electromagnet in which the current is switched on or off, will also induce a voltage in the coil as the field lines 'grow' or 'collapse'.

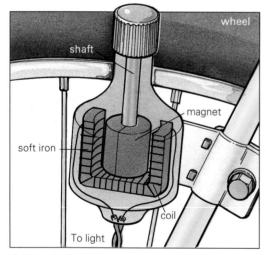

A bicycle dynamo.

Generating electricity

Electromagnetic induction is the principle behind the generation of electricity. A generator (or dynamo) works by rotating a coil of wire in a magnetic field.

In a bicycle dynamo, however, a magnet rotates inside a coil. This has the same effect — remember it is the relative motion of a conductor and a magnetic field that creates the voltage.

Mains electricity is produced at power stations by large generators called **alternators** (see 3.6) which give an alternating current (a.c.) output (see 2.1). The magnetic field is created by electromagnets, and it is these which rotate while the 'induction' coils remain stationary. The rate of rotation (in the UK) is 50 revolutions per second, which gives an output with a frequency of 50 Hz.

How does a generator work?

A simple **d.c. generator** has a similar structure to a simple motor (see 3.3), and, in fact, a motor can be used to generate a current. Every half revolution the commutator reverses the connection with the brushes. If the brushes are connected to an external circuit, this produces a current which is always in the same direction, although not smooth — it varies between maximum and zero as the coil rotates.

An **a.c. generator** does not use a commutator – each end of the coil is connected to a full ring called a **slip ring**, which is always in contact with the same brush. The current generated in an external circuit is constantly changing from a maximum value in one direction to zero to a maximum value in the reverse direction to zero and then back to the beginning of the cycle.

The frequency of the alternating current, or the number of cycles per second, is equal to the rate of revolution of the coil. You can verify this by connecting a simple a.c. generator to an oscilloscope. What other effect do you see?

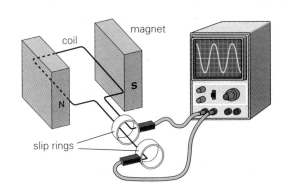
A simple a.c. generator and its output.

Other applications of electromagnetic induction

A car needs a high voltage to operate the spark plugs (about 10kV). The supply is a 12V battery. Two coils – a primary coil connected to the battery, and a secondary ('induction') coil connected to the spark circuit – are wound on an iron core. The high voltage is induced in the secondary coil by breaking the primary circuit, thus causing the magnetic field of the primary coil to collapse rapidly.

Tape recordings store signals on magnetic tape. Electromagnetic induction enables tape recordings to be played. When a tape is played the magnetic particles on the tape pass over the playback heads which contain coils. Small voltages are induced in these coils, creating the electrical signals from which sound is reproduced in a loudspeaker.

In a moving-coil microphone, sound waves cause a coil to vibrate in a magnetic field. Varying voltages are induced in the coil, which follow the variation in frequency and amplitude of the sound waves.

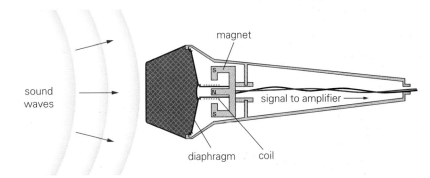
A microphone converts sound signals into electrical signals.

QUESTIONS

1 What are the factors which affect the *size* of an induced voltage across a coil of wire when a magnet is moved near it? What factors affect the *direction* of the induced voltage?

2 The construction of moving-coil microphones and moving-coil loudspeakers (see 3.3) are very similar. Describe the energy conversions in each, and identify the main difference in their construction.

3 How is the frequency of the mains a.c. supply determined?

4 When a magnet is moved towards a coil of wire connected in a circuit, an induced current will flow in the circuit. This produces a magnetic field which *opposes the motion of the magnet*. Why would it not be possible for the induced current to flow in the opposite direction?

3.5 Transformers

If an a.c. supply is connected to an electromagnet, the magnetic field is constantly changing as the current increases, decreases and changes direction. A coil within this changing magnetic field will have a voltage induced in it, which will change in the same way as the a.c. supply. This effect is used in the **transformer**.

Transformers enable us to change the voltage of an a.c. supply. The induced voltage may be larger or smaller than the supply voltage, depending on the number of turns on the coil. If the induced (output) voltage is larger, the transformer is called a **step-up** transformer. If it is smaller, the transformer is called a **step-down** transformer.

The power supply to a word processor contains a transformer.

Although you may not see them, there are transformers inside many everyday electrical appliances. Any mains electronic devices, or those which can be run from mains *or* from batteries, will have a transformer. It may be part of the **mains adaptor** which is plugged into a socket, or it may be inside the device where the mains supply enters. A television set has both step-up and step-down transformers – the step-down transformer is needed to reduce the 230 V a.c. mains voltage to about 12 V for the electronic circuits, while the step-up transformer provides the 15 kV needed to accelerate the electrons in the picture tube.

The fact that transformers can change a.c. voltages to any value required is the reason that mains electricity supplies are a.c. Transformers will not work with d.c., and there is no equivalent device to change d.c. voltages without losing a lot of energy in the process.

Some appliances, such as those with electronic circuits, need a d.c. supply to work – the a.c. needs to be converted to d.c., or **rectified**, using a circuit containing a diode.

Construction of a transformer

A transformer consists of two separate coils wrapped round the same soft iron core. An alternating input to one coil will induce an alternating voltage in the other. The size of the voltage induced depends on the size of the input voltage, and the ratio of turns in the two coils. The input voltage is applied to the **primary coil**, and the output voltage is induced in the **secondary coil**. If there are more turns in the secondary than in the primary, then the voltage will be increased. This is a step-up transformer. If there are fewer turns in the secondary than in the primary, the voltage will be decreased. This is a step-down transformer.

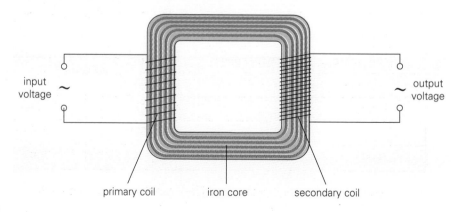

A simple step-up transformer.

The relationship between the primary and secondary voltages, and the number of turns in the coils, is given by

$$\frac{\text{primary (input) voltage}}{\text{secondary (output) voltage}} = \frac{\text{number of turns in primary coil}}{\text{number of turns in secondary coil}}$$

So output voltage = $\dfrac{\text{input voltage} \times \text{number of turns in secondary coil}}{\text{number of turns in primary coil}}$

In symbols: $\dfrac{V_p}{V_s} = \dfrac{N_p}{N_s}$ and $V_s = \dfrac{V_p \times N_s}{N_p}$

Example: To calculate the output voltage of this step-down transformer, use the above relationship:

$$V_s = \frac{230\text{V} \times 100}{2000} = 11.5\text{V}$$

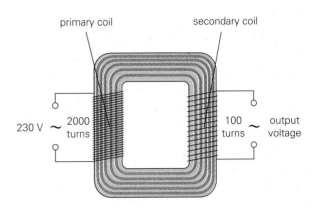

Transformers change the voltage of a.c. supplies with very little loss of electrical energy. Modern transformers are better than 99% efficient. This means that less than 1% of the electrical energy input is lost. The energy losses that do occur are as a result of heating in the coils of the transformer, and the induction of voltages in the iron core. These lead to stray currents, called **eddy currents**, which dissipate energy as heat. These losses are reduced by laminating the iron core – it is built up of thin layers of iron separated by insulating layers. This has little effect on the magnetic properties, but prevents currents flowing across the laminations.

A transformer, showing the laminated core.

QUESTIONS

1 Explain the difference between a 'step-up' and a 'step-down' transformer.

2 Make a list of household appliances in which a transformer is used.

3 What is the relationship between the number of turns in the coils of a transformer and the input and output voltages? Copy the table below with details of transformers and fill in the gaps.

Primary turns	Secondary turns	Input voltage	Output voltage
400	20	230 V	
200		230 V	11.5 kV
	2200	25 kV	275 kV
1000	400		200 V

4 How is energy lost from a transformer? How can this loss be reduced?

3.6 Generation and transmission of electricity

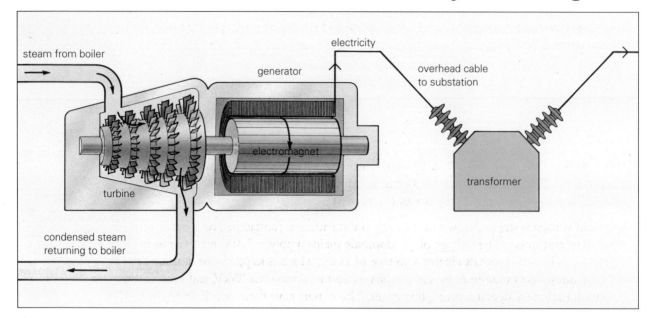

Power station alternators generate a.c. at a voltage of 25 kV, and can produce a current of up to 20 kA. The rotating electromagnets are driven by turbines which require steam at very high pressures. The steam is heated by burning coal, gas or oil, or by a nuclear reactor.

The power generated by an alternator can be calculated from

$$\begin{aligned} P &= I \times V \\ &= 20\,000\,\text{A} \times 25\,000\,\text{V} \\ &= 500\,000\,000\,\text{W} \\ &= 500\,\text{MW} \end{aligned}$$

The power is fed into a network of cables known as the **National Grid**, which allows the power to be switched to areas where it is needed.

Use of transformers

Transformers are essential in the transmission of electricity from power stations to customers. The transmission involves transferring the electrical energy over long distances. There is always heat produced when a current passes through a wire – the energy dissipated per second is equal to I^2R (see 1.9), where I is the current and R is the resistance of the wire. The heat generated in long-distance cables can therefore be reduced by using thicker wires in the cables, but this will increase the cost. Thinner wires are cheaper and the cables will be lighter, so that supporting them will also be cheaper.

These cables, supported by pylons, form part of the National Grid.

The current can be decreased, without reducing the electrical power transferred, by increasing the voltage. The 25 kV output from power station alternators is stepped up to 275 kV or 400 kV for transmission.

If an alternator has an output of 500 MW at 25 kV, and the voltage is stepped up to 400 kV, the current will be:

$$I = \frac{P}{V}$$
$$= \frac{500\,000\,000\,\text{W}}{400\,000\,\text{V}}$$
$$= 1250\,\text{A} = 1.25\,\text{kA}$$

Although the power transmitted at the high voltage is the same, the current has been reduced from 20 kA to 1.25 kA. This means that the energy lost as heat will be much reduced, as this depends on the square of the current.

The Grid voltage is stepped down in stages by transformers at substations, to the voltage supplied to customers. The voltage of the domestic mains supply is 230 V, but large users such as factories and hospitals receive a voltage of 11 kV, which is stepped down inside the building. Some users require even higher voltages, some railways use 25 kV, and some heavy industrial machinery operates on high voltages. These users have their own links to the Grid.

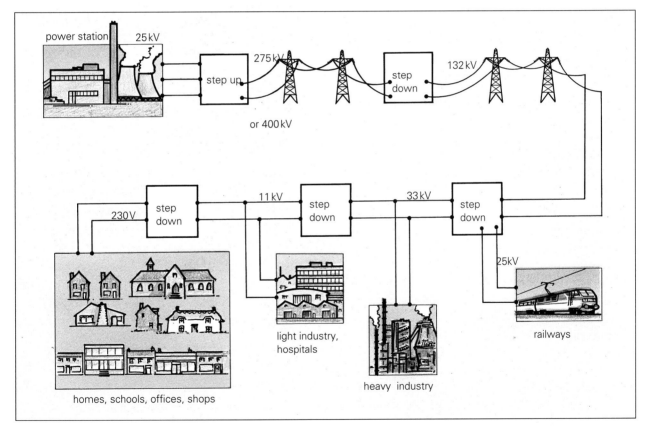

The National Grid transmission network.

QUESTIONS

1 What sources of energy are used to generate electricity?

2 Why is mains electricity generated as a.c.?

3 Explain why transformers are used to step up voltages for transmission of electricity over long distances.

SECTION A: QUESTIONS

1. A household lamp uses fine tungsten wire as the filament. What are the properties of tungsten which make it suitable for this? The bulb contains argon gas at low pressure. Why is this used in place of air?

2. A power station generates an alternating current supply at a voltage of 25 kV. The voltage is stepped up by a transformer to 400 kV for connection to the National Grid. The power output of the station is 125 MW.
   a Calculate the current flowing from the generator.
   b If the primary coil of the transformer has 4000 turns, calculate how many turns are required in the secondary coil.
   c Assuming that the transformer is 100% efficient, calculate the current flowing from the output coil.

3. The diagrams show two methods of connecting together four 6Ω resistors with a 12 V power supply.
   a Calculate the current in *one* of the resistors in method A.
   b Calculate the *total* current taken from the battery in method A.
   c Calculate the *total* current taken from the battery in method B.
   d Calculate the *total* power developed in the four resistors by each of the methods of connection.

4a What materials are attracted by a magnet? Describe one practical use for a permanent magnet.
   b Describe how you would make an electromagnet. How could you make it stronger?
   c Describe one practical use for an electromagnet, excluding any for which a permanent magnet could be used. Explain why you could not use a permanent magnet.

5. Explain why a transformer cannot be used with a d.c. supply. Is it possible to change the voltage of a d.c. supply? If so, explain how it can be done.

6. Ring mains are used to connect sockets to the electricity supply in homes.
   a Describe with a diagram how a ring main is connected.
   b Explain the advantages of using ring mains.
   c Would you connect an electric cooker, which may take a current of 40 A, to a ring main? Explain your answer.

7. Diagram A shows a positively charged small ball hung on insulating thread. Diagrams B, C and D show what happens when three charged strips, X, Y and Z, are brought near to the ball.
   a What does this tell you about the charge on each of the three strips?
   b How do you think the strips were given their charge?
   c Why is the ball hung on insulating thread?
   d What do you think would happen if strip X touched the ball?

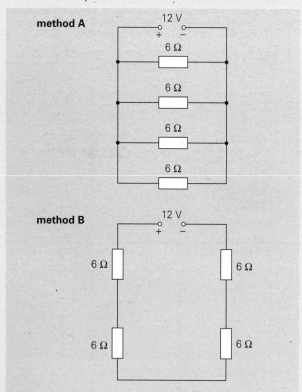

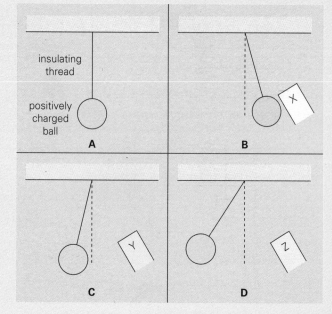

8a Name an electrical appliance which converts electrical energy to: **i** heat energy, **ii** sound energy, **iii** kinetic energy.

b What will be the total cost if two 60 W lamps and two 40 W lamps are switched on for six hours a day while the occupants of a house take a two-week holiday? The cost of electricity is 7p per unit.

9 Sketch graphs of current I against voltage V, with labelled axes and scales, to show what you would expect when the voltage is varied from 0 to 12 V across:

a a 0.5 m length of resistance wire, with resistance 12 Ω per metre

b a 12 V 24 W filament lamp.

Explain why the graphs are different.

10 A simple electric motor uses permanent magnets to provide the magnetic field, but a commercial motor is more likely to use electromagnets. One reason for this is that the motor can then be used with either an a.c. or a d.c. supply. Explain why this is possible. Describe one other advantage of using electromagnets in a motor.

11 A battery charger is used to recharge a 12 V car battery which is completely exhausted. The charging current is 2 A, and the battery is fully charged after 7 hours.

a Calculate how much charge, measured in coulombs, flows into the battery.

b If the same amount of charge flowing from the battery will exhaust it, calculate how long the battery will last when it is used to supply the starter motor, which takes 120 A.

c Calculate how long the battery would last if the headlights (two 12 V 36 W lamps) and sidelights (four 12 V 6 W lamps) were left on.

12a The diagram shows apparatus to be used to plate a metal spoon with silver.

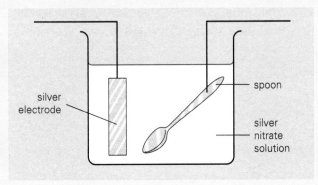

The apparatus, which consists of a silver electrode and silver nitrate solution, is connected to a power supply. **i** Which terminal of the power supply should be connected to the spoon? **ii** What name is given to this electrode?

b A current of 0.1 A flows through the apparatus for 5 minutes to plate the spoon. **i** How much charge has passed through the solution? **ii** How could the quantity of silver plating on the spoon be increased?

c The silver plating on the spoon is uneven, with more silver on the side nearest to the silver electrode. How could the apparatus be changed to provide an even plating of silver over the surface of the spoon?

13a Diagram A shows a step down transformer connected to a 230 V 50Hz a.c. supply. There are 460 turns in the primary coil, and 24 in the secondary coil. What is the output voltage of the transformer?

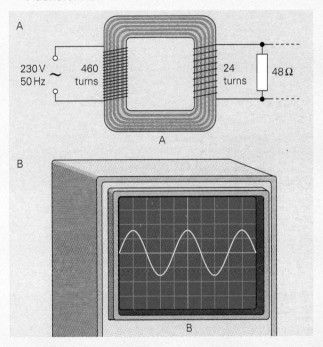

b A device of resistance 48 Ω is connected to the output of the transformer. **i** What current flows through the device? **ii** What is the power taken by the device?

c Diagram B shows the screen of an oscilloscope (CRO) connected across the device. Sketch the display and show on your sketch the traces if the secondary coil of the transformer had **i** 12 turns, **ii** 48 turns.

d If the grid is divided into centimetre squares, how much time is represented by each centimetre on the horizontal axis of the grid?

CHAPTER 4: FORCES AND MOVEMENT

4.1 Force

How do we know when a force is acting?

Many of the great scientists in the seventeenth century spent time developing ideas about **force**. Isaac Newton is the most famous. He is remembered for his laws of motion and for developing the theory of gravitational force, or **gravity**. Gravitational forces act over very large distances; other forces, like those holding the nucleus of an atom together, act only over very small distances. Forces may be very large or very small. No matter how large a force may be, you can never see it. What you see are the *effects* of the force – the apple falling from the tree, the Moon orbiting the Earth, a damaged car after a road accident, even just a door being opened. Forces make things happen to objects, by altering their movement, or by changing their shape. Pushing, pulling, turning, squeezing, stretching and twisting all involve forces.

Isaac Newton, 1642–1727. Experiment or accident?

Weight is a force

Weight is a force. Like all other types of force it is measured in **newtons** (N). The weight of an object is simply the force that acts on it because of the planet's gravity. It always acts in a downwards direction (towards the centre of the planet). The greater the mass of an object, the greater the pull of gravity, and so the greater its weight. Mass is measured in **kilograms** (kg). The downwards force due to the Earth's gravity is 9.8 N (roughly 10 N) for each kilogram of mass. This is called its **gravitational field strength**. A mass of 10 kg therefore weighs about 100 N.

Weight and mass are related by the formula:

weight (N) = mass (kg) × gravitational field strength (N/kg)

or in symbols:

$W = m \times g$ or $W = m\,g$

On the Earth $g = 9.8$ N/kg.

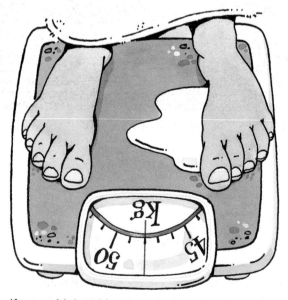

If you said that this person weighed 48 kg, strictly speaking you would be wrong. It is their mass which is 48 kg, making their weight about 480 N.

Equal and opposite forces

One of Newton's laws says that *whenever two bodies interact, the forces they exert on one another are always equal and opposite*. These photographs show this law 'in action'.

Consider the fish hanging from the spring balance. The fish exerts a downward pull equal to its weight on the balance. At the same time the balance provides an upward force of equal size on the fish. If you remove this force, the fish would fall.

The effects of the forces may be more apparent in the case of the tennis racquet and ball. The force of the racquet on the ball stops its motion and propels it away again; the force of the ball on the racquet deforms the racquet strings.

The spring balance pulls upwards on the fish with a force equal to the weight of the fish.

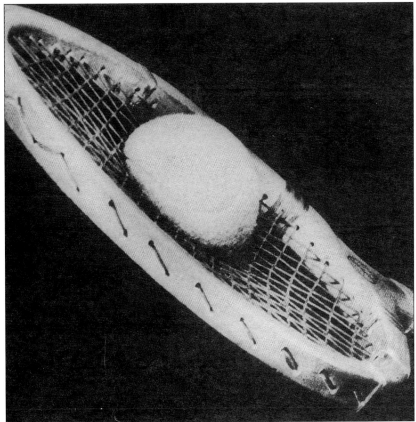

As the racquet hits the ball, an equal and opposite force affects the racquet.

QUESTIONS

1. What is the weight of a bag of shopping of total mass **a** 6 kg, **b** 5 kg?

2. What is the mass of a person of weight **a** 700 N, **b** 650 N?

3. What forces are acting on a book resting on a table?

4. What would happen to **a** your mass, **b** your weight, if the Earth's gravitational field strength were to double?

4.2 Resultant force

Direction makes a difference

The effect that a force has on an object depends not only on the *size* of the force but also on its *direction*. Common sense tells you that to raise a glass, you need to apply a force in an upwards rather than in a sideways direction.

When talking about forces we should therefore specify not only their size, but also their direction. The direction might be described as 'to the left' or 'to the right', or as 'acting in an opposite direction'. In more complicated situations, the direction may be given as a compass bearing.

Quantities such as forces which have a direction associated with them are called **vector** quantities. In this chapter, you will come across three vector quantities: force, velocity and acceleration. In each case, knowing their direction is as important as knowing their size (or **magnitude**). Quantities such as volume, which don't have a direction associated with them, are called **scalar** quantities.

In diagrams, a vector quantity can be represented by an arrowed line in the direction of the vector and whose length is proportional to the magnitude of the vector.

An upward force is needed to raise a glass.

Adding forces

There are usually several forces acting on an object at any time. Adding forces together is more complicated than, for example, adding volumes of water, because the direction of each force needs to be taken into account.

If you add $50\,cm^3$ of water to $100\,cm^3$ of water, you get $150\,cm^3$ of water. But if you add a force of 4 N to a force of 3 N, you could end up with a force as great as 7 N, or as little as 1 N.

Two forces acting in the same direction add to give a force acting in the same direction, whose magnitude is equal to the sum of the two magnitudes.

Two forces acting in opposite directions add to give a force whose direction is the same as that of the larger force, whose magnitude is equal to the difference of the two magnitudes.

If the two forces don't act along the same straight line, they add to give force in a different direction, whose magnitude is less than their sum (but greater than their difference).

When all the forces acting on an object are added together, their vector sum is called the **resultant** force.

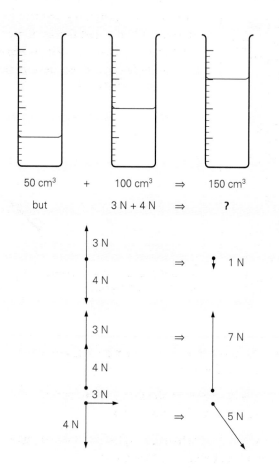

Vector drawings can help when adding forces together.

No resultant force – no change

Objects which are stationary have forces acting on them, but their resultant is always zero. The forces 'balance each other out'. For example, the fish hanging from the spring balance (see 4.1) has no resultant force on it – its weight acting downwards is equal to the supporting force upwards. Similarly, an object resting on a surface has its weight balanced by an upward force from the surface.

An object travelling at a steady speed in a straight line, such as a car, also has no resultant force acting on it. If the thrust of the engine were increased there would be a resultant force in the direction of motion and the car would start to speed up. If the thrust were decreased instead, there would be a resultant force in the opposite direction and the car would start to slow down.

The four forces balance each other out so the car continues at a steady speed.

Direction of resultant force	acting on	Result
zero	stationary	no change — still stationary
zero	steady speed	no change — same speed and direction
→	stationary	speed increases
→ (larger)	steady speed	speed increases
←	steady speed	slows down to a stop then speeds up in direction of force

In general, if the forces acting on an object don't add up to zero, then the speed and/or the direction of movement of the object will change, and continue to change while a resultant force continues to act.

Isaac Newton summed all this up when he formulated his first law of motion. When stated simply, it says that *things stay at rest or continue moving in a straight line at a steady speed unless acted on by a resultant force.*

A resultant force will change the motion of an object.

QUESTIONS

1. In which of these situations is a resultant force acting? In each case, justify your answer.
   a. a parachutist falling at a steady speed through the air
   b. a plate on a table
   c. a car pulling away at traffic lights
   d. a car travelling at a steady speed in a straight line
   e. a car going around a bend at a steady speed
   f. a car slowing down

2. Draw diagrams to show the forces acting in parts **a** to **d** of question **1**.

3. Explain why, when adding forces, their direction needs to be taken into account.

4.3 Speed

Speed is measured in **metres per second** (m/s). If you know the speed of something, you can work out how far it will go in a particular time. If a car has a steady speed of 30 m/s, it will travel 30 metres in one second, and will travel a further 30 metres in each second that follows.

Working out speed

The speed of a steadily moving object can be worked out using the formula:

$$\text{speed (m/s)} = \frac{\text{distance travelled (m)}}{\text{time taken (s)}}$$

or, in symbols: $s = \dfrac{d}{t}$

So, if you travel the same distance in half the time, you will have doubled your speed.

A graph of distance against time is called a **distance–time graph**. The shape of the graph enables you to see at a glance how the object is moving. The steeper the slope, the greater the speed.

On most journeys the speed is not constant. It is sometimes useful to talk of the **average speed**:

$$\text{average speed (m/s)} = \frac{\text{total distance travelled (m)}}{\text{time taken (s)}}$$

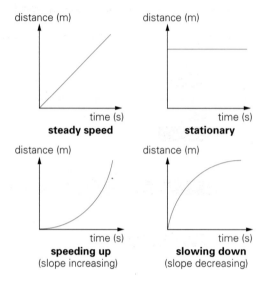

The slope of a distance–time graph is equal to the speed of the object.

How fast?

Speed of light	300 000 000 m/s
Speed of sound	300 m/s
Car on motorway	30 m/s
Person jogging	3 m/s
Child crawling	0.3 m/s
Snail	0.03 m/s

A stopwatch, a ruler and a piece of string can be used to measure the average speed of a snail. Try it!

Speed and road safety

It takes an alert driver about 0.7 s to apply the brakes in response to seeing a hazard. This time is known as the thinking time. The thinking time does not depend on the car's speed. A car travelling at 60 mph will travel twice as far in 0.7 s as one travelling at 30 mph so the thinking *distance* is twice as much.

The braking distance is the distance the car will travel between the time the brakes are applied, and the time that if finally stops. Doubling the speed *quadruples* the braking distance.

The separation of many of these vehicles is less than that recommended by the Highway Code.

The greater the speed of the car:

- the greater the braking force needed to stop it in a certain time, or
- the longer the time needed to stop it with a certain (e.g. maximum) braking force, and so the further it travels before it stops.

When roads are wet or icy the braking distance is greater.

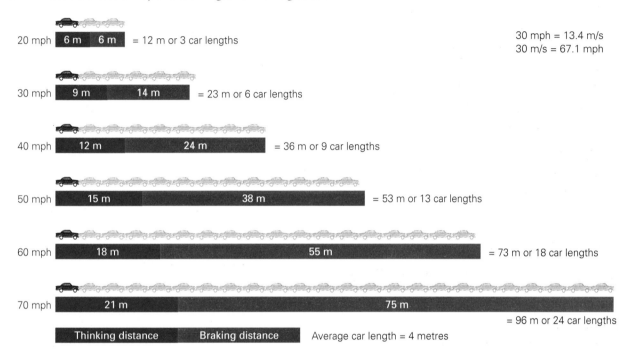

30 mph = 13.4 m/s
30 m/s = 67.1 mph

The stopping distance is the thinking distance plus the braking distance.

What is velocity?

If you know the speed of an object and its direction of travel, then you know its **velocity**. The velocity of something is its speed in a stated direction – it is a vector quantity, like force (see 4.2). Whenever the direction changes, so does the velocity. Like speed, it is measured in metres per second (m/s).

In simple situations, an object travelling in a straight line in one direction can be described as having a positive velocity whereas another travelling towards it in the opposite direction would be described as having a negative velocity, e.g. – 30 m/s. The minus sign indicates that it is travelling in the opposite direction.

Velocity is worked out using the formula:

$$\text{velocity (m/s)} = \frac{\text{distance travelled in a given direction (m)}}{\text{time taken (s)}}$$

QUESTIONS

1. Calculate the average speed of a car if it travels:
   a 1000 m in 50 s;
   b 4800 m in 10 minutes.

2. If a lorry travels at an average speed of 30 m/s, how long will it take to travel
   a 10 km; b 33 km?

3. Use the chart of shortest stopping distances to plot a graph of
   a braking distance against speed,
   b overall stopping distance against speed.

4. A car travelling at a steady speed travels 1000 m in 90 s.
   a Plot a graph of distance against time.
   b From your graph, read off the distance gone after i 20 s, ii 40 s, iii 60 s.
   c Calculate the slope of the graph, and hence the car's speed.

4.4 Acceleration

Acceleration is the rate at which the velocity of something is changing. It is measured in **metres-per-second per second**, which is written as m/s². It tells you how much the velocity will change each second. If an initially stationary car drove off in a straight line with a steady acceleration of 2 m/s², its velocity would increase by 2 m/s every second. So after 1 second its velocity would be 2 m/s, after 2 s it would be 4 m/s and so on.

Working out acceleration

Acceleration can be calculated by using the formula:

$$\text{(average) acceleration (m/s}^2) = \frac{\text{change in velocity (m/s)}}{\text{time taken for the change (s)}}$$

or, in symbols: $a = \dfrac{v - u}{t}$

where u is the velocity at the beginning of the time interval and v is the velocity at the end of the time interval.

When an object is slowing down, the change in velocity is negative (because v is less than u), and so the acceleration is negative. This is sometimes called a *deceleration*.

Velocity–time graphs

The shape of a velocity–time graph enables you to see at a glance how the velocity is changing. The greater the slope, the greater the acceleration.

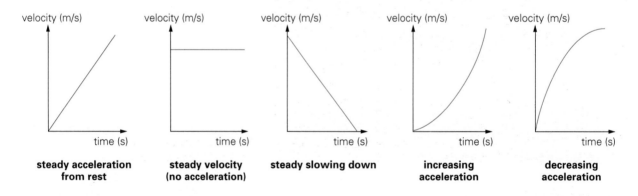

steady acceleration from rest   **steady velocity (no acceleration)**   **steady slowing down**   **increasing acceleration**   **decreasing acceleration**

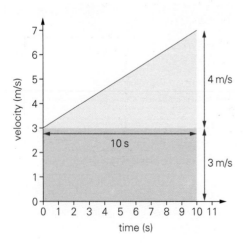

The acceleration at any point in a journey can be calculated by measuring the slope of a velocity–time graph. In effect this is the same as applying the formula above. For the object whose velocity–time graph is shown on the left:

$$\text{acceleration} = \text{slope} = \frac{4 \text{ m/s}}{10 \text{ s}} = 0.4 \text{ m/s}^2$$

The area under the graph is equal to the distance travelled. So the distance travelled in the first 10 seconds is equal to the area of the darker blue rectangle plus the pale blue triangle:

$$\text{distance travelled} = (10 \text{ s} \times 3 \text{ m/s}) + (\tfrac{1}{2} \times 10 \text{ s} \times 4 \text{ m/s})$$

$$= 30 \text{ m} + 20 \text{ m} = 50 \text{ m}$$

How much acceleration?

Rocket blasting into space	100 m/s²
Object falling to the floor	10 m/s²
Train pulling out of a station	1 m/s²
Ferry moving away from its moorings	0.1 m/s²

In general, the acceleration of an object depends on:

- the size of the force causing the acceleration
- the mass of the object.

When a resultant force acts on an object, it speeds up in the direction of the force (see 4.2). The greater the resultant force due to the thrust of a car's engine, for example, the greater the acceleration of the car. A heavier car, however, will have a smaller acceleration for the same thrust. This is why dragsters (cars which need to be able to accelerate quickly) have their mass made as small as possible, and engines capable of producing large forces.

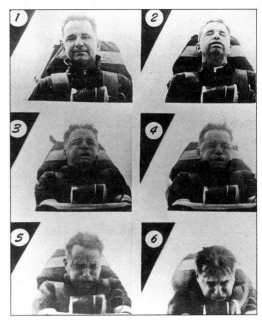

Large forces are exerted on an astronaut as a rocket accelerates.

Dragsters are designed for maximum acceleration.

QUESTIONS

Questions 1 and 2 represent simplified versions of real situations. It is assumed that the acceleration is constant over the time period considered. In reality it is unlikely to be the case.

1. Calculate the acceleration of a car if its velocity:
   a increases from 0 to 30 m/s in 12 seconds
   b increases from 10 m/s to 25 m/s in 5 seconds
   c decreases from 20 m/s to 10 m/s in 10 seconds.

2. A train accelerates at a steady rate from rest for 30 seconds, by which time it has reached a velocity of 15 m/s. Draw a velocity–time graph, and use it to calculate:
   a the train's acceleration, and
   b the distance travelled in the first 30 seconds of its journey.

3. Describe (with terms like 'speeds up', 'slows down', 'more quickly', etc.) the motion of the vehicle whose velocity–time graph is shown here.

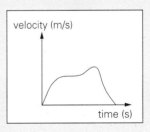

4.5 Force and acceleration

The relationship between acceleration, force and mass is given by the formula:

$$\text{acceleration (m/s}^2\text{)} = \frac{\text{force (N)}}{\text{mass (kg)}}$$

or, in symbols:

$$a = \frac{F}{m}$$

When no resultant force is acting, there is no acceleration, and therefore no change in velocity. A resultant force produces an acceleration proportional to the force. Doubling the force applied will double the acceleration, whereas doubling the mass of the object for the same force will halve the acceleration.

The formula can be rearranged to find the force needed to give an object a particular acceleration:

$$\text{force (N)} = \text{mass (kg)} \times \text{acceleration (m/s}^2\text{)}$$

or, in symbols:

$$F = m \times a \text{ or } F = m\,a$$

The size of a newton

From the formula above it can be seen that one newton (1 N) is the force needed to give a mass of 1 kg an acceleration of $1\,\text{m/s}^2$.

Huge amounts of fuel are used each second to provide the thrust needed to accelerate a rocket.

Weight and acceleration

A freely falling object has only one force acting on it: its weight, in a downwards direction. Its weight causes it to accelerate at a steady rate. Because

$$a = \frac{F}{m}$$

and $F = W = m \times g$

a is numerically equal to g, the gravitational field strength.

In an evacuated tube (one where the air has been removed), objects will fall with the same acceleration.

Near the Earth's surface, all freely falling objects therefore have the same acceleration: $9.8 \, m/s^2$. This acceleration is known as the acceleration due to gravity.

On the Moon, the acceleration of a freely falling object is one-sixth of its value on the Earth, because the Moon's gravitational field strength is one-sixth of the Earth's.

The Moon has no atmosphere, so any object which is dropped will fall freely; on the Earth, the atmosphere prevents this from happening. It has a greater effect on the motion of some objects than of others. The effect on a coin is minimal, unlike its effect on a feather.

Circular motion

Things moving in circles are accelerating. For example, although each person on the roundabout is spinning at a steady speed, their velocities are continuously changing because their direction of travel is continuously changing. Because their velocity is changing, they are in fact accelerating. The force causing each person to accelerate is applied through the chains connected to their chair. The direction of this force changes as the roundabout spins. If the chains broke (so that the force was no longer applied), each person would fly off in a straight line along a tangent. Although their weight would now make them accelerate towards the ground, their horizontal velocity would remain unchanged.

QUESTIONS

1a Calculate the acceleration of a boat of mass $10^8 \, kg$ if the (constant) resultant force (the thrust minus the drag) acting on it is $8 \times 10^6 \, N$.

  b How long will it take to reach a velocity of $5 \, m/s$?

2 If the mass of an astronaut is $65 \, kg$, calculate her weight on the Earth and what it would be on the Moon.

3 Calculate the acceleration of a freely falling coin and a feather on the Moon.

4a Calculate the size of the resultant force (the thrust minus the drag) needed to give a car of mass $1100 \, kg$ a steady acceleration of $3.0 \, m/s^2$.

  b Calculate the velocity the car would have after 5 seconds, if it started off with a velocity of $15 \, m/s$.

5 If the thrust produced by a rocket is constant, how will its acceleration change with time? Explain your answer.

4.6 Friction

Friction is a force linked to movement: it slows things down, or stops them from moving. *The direction of the frictional force is always opposite to the direction in which the object is moving or trying to move.* Friction arises because of the roughness of the two surfaces which are moving or trying to move past one another – the roughness provides a resistance to the movement. The greater the resistance to the movement, the greater the frictional force. Frictional forces hold nails and screws in place, they stop things like chairs and tables from sliding around. Frictional forces are what give us **grip**!

Frictional forces generate heat between moving surfaces (try rubbing your hands together) and can at the same time cause surface wear.

Increasing friction

Sometimes it is important to maximise the amount of friction between two surfaces in order to increase the amount of grip available. Car brakes and tyres are designed with this in mind. If a road surface is wet or icy, the grip is reduced. Under these conditions a car will not only need a greater stopping distance, but if the brakes are suddenly jammed on, the car is much more likely to skid (and go out of control).

The amount of grip which is lost in wet weather depends on the tread pattern of the tyre.

Reducing friction

In some situations, such as in machinery, it is important to try to reduce the amount of friction so that there is less wear, less heat generated, less noise, and therefore greater efficiency. This can be achieved through good design, using polished moving parts, and by using **lubricants**.

The job of a lubricant is to make surfaces more slippery. Although many lubricants (including oils) are liquids, others such as graphite powder are solids. A well-chosen lubricant will:

- reduce the amount of friction
- stay on the parts it is lubricating
- be chemically inactive under the conditions in which it is being used.

Lubricant 'fills in' irregular surfaces, separating them and reducing energy loss.

Drag

When an object moves through a fluid (a liquid or gas), the opposing frictional forces are usually called **drag**. The amount of drag depends on the shape of the object, and so can be reduced by good design.

'Streamlining' has helped improve the energy efficiency of modern cars.

Faster and faster?

The faster an object moves through a fluid, the greater the amount of friction or drag which opposes its motion. This is why, despite the forward thrust of a car's engine, a car does not keep getting faster and faster.

If a ball bearing is dropped into wallpaper paste, it experiences forces in both an upward and downward direction. At the start, these forces add to give a resultant force in a downwards direction. As the ball bearing falls, it therefore accelerates and its velocity increases. The upward force due to the drag gets larger, and consequently the resultant force downwards gets smaller.

Eventually, a state is reached where the resultant force is zero because the upward and the downward forces balance each other out. When this happens, the ball bearing stops accelerating and continues at a steady speed until it hits the bottom of the tube. This steady speed is known as its **terminal velocity**.

The size of the terminal velocity depends on the mass, volume and shape of the object, and on the type of fluid it is falling through. The greater the frictional resistance, the lower the terminal velocity (because the steady state is reached sooner).

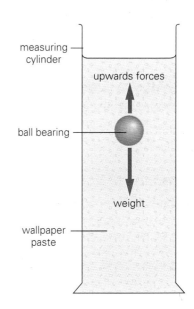

A ball bearing falling through wallpaper paste has both upwards and downwards forces acting on it.

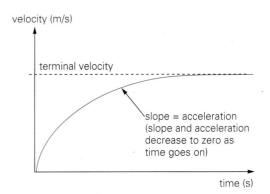

A freely falling object reaches a terminal velocity.

A falling cat can increase the amount of drag by spreading itself to make a less streamlined shape.

QUESTIONS

1. Explain why racing cars have their tyres changed when it rains heavily.

2a. Make a list of circumstances in which friction would cause wear, and is reduced by using lubricants.
  b. Make a list of circumstances in which friction is deliberately increased.

3. Explain, with the help of sketch graphs, why a person jumping from an aeroplane with a parachute can land unharmed, whereas a person jumping without one would probably be killed.

4.7 Turning effects

Forces can make objects rotate around a pivot. The turning effect of a force is called its **moment**. The word **torque** is also used. Spanners and screwdrivers are designed to make use of the turning effect of forces.

How big is a moment?

The turning effect due to the weight of the girl on the right in the photo is less than the turning effect due to the weight of the girl on the left. As a result, the seesaw is rotating in an anticlockwise direction.

The size of each moment is equal to the force multiplied by the perpendicular distance from the line of action of the force to the pivot (the point about which the seesaw is turning).

This shows that, for a given force, the turning effect is greater the further from the pivot the force is applied. You use this fact every day when you open or close a door. It is why spanners have long handles, and is the principle behind all sorts of levers.

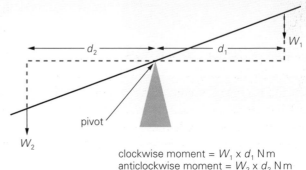

clockwise moment = $W_1 \times d_1$ N m
anticlockwise moment = $W_2 \times d_2$ N m

Calculating moments.

Equilibrium

For the motion of an object to be unaffected by the forces acting on it (for example not to move at all), two things have to happen at the same time:

- the resultant force acting on it must be zero (see 4.2), and
- all the clockwise moments about any point must be exactly balanced by all the anticlockwise moments about the same point.

The object is then in **equilibrium**.

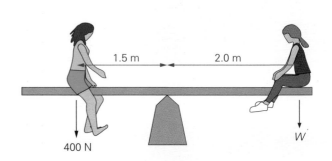

Example: In the diagram, the two girls sitting on the seesaw are balanced. The seesaw is not moving, so the clockwise moment of the girl on the right must be exactly balanced by the anticlockwise moment of the girl on the left. If the girl on the left weighs 400 N, then because we know how far each of the girls is from the centre, we can calculate the weight W of the girl on the right.

anticlockwise moment about pivot = 400 N × 1.5 m

clockwise moment about pivot = W × 2.0 m

Therefore

$W \times 2.0\,\text{m} = 400\,\text{N} \times 1.5\,\text{m}$

Rearranging gives

$$W = \frac{400\,\text{N} \times 1.5\,\text{m}}{2.0\,\text{m}} = 300\,\text{N}$$

Some applications of turning effects

In some machinery, it is important to do up the nuts and bolts to just the right degree of tightness. Too much, and there is a danger of the nut or bolt failing mechanically. Too little, and the nut may become loose because of vibrations. A 'torque wrench' is a calibrated spanner (one with a scale) with a handle that flexes. The greater the torque applied, the greater the flexing and the greater the reading.

A specified torque about to be applied.

When using a screwdriver, it is the force of the hand on the handle that produces the turning effect. The thicker the handle of a screwdriver, the greater the torque that can be produced. In practice, the diameter of a screwdriver handle is matched to the size of the screw-head it is designed to tighten. This reduces the risk of damage to the screws or the material that they are being screwed into.

Wheelbarrows are used to move heavy loads more easily. Once the handles have been lifted, the moments about the wheel (the pivot) in one direction due to the weight of the load and the barrow are exactly balanced by the moment in the opposite direction due to the lifting force. Since the distance to the handles is greater than the distance to the load, the size of the lifting force needed is less than the load.

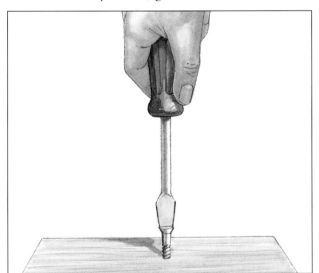

The amount of torque is controlled by diameter of the handle and the force applied.

The magnitude of the lifting force E is less than the weight of the load.

QUESTIONS

1. Where on a seesaw must a child (weight 400N) sit to exactly balance a second child (weight 500N) who is sitting 1.5m from its centre?

2. A child (weight 500N) is sitting 2.0m from the end of the seesaw. A second child (weight 400N) is sitting at the other end, the same distance from the centre. Where must a third child of weight 400N sit in order to balance the seesaw?

3. Explain why the further from the hinged edge a force is applied, the easier it is to push open a door.

CHAPTER 5: FORCES, SHAPE AND PRESSURE

5.1 Solids, liquids and gases

How do we decide whether to classify something as a solid, a liquid or a gas? Shape and volume are two of the factors which we unconsciously tend to take into account.

Solids have a shape of their own, whereas liquids and gases have a shape which varies according to the shape of their container. When transferred to a larger container, the volume of a liquid will stay the same, whereas a gas will always expand to fill the container.

It is possible to explain these and many other properties of solids, liquids and gases in terms of what is happening to the particles that make them up.

We divide substances into solids, liquids and gases almost without thinking.

Molecules

The smallest piece of matter that can exist on its own under normal conditions is a **molecule**. Molecules are so small that they can't be seen, even with the most powerful microscope. A large glass of water contains around 10^{25} molecules!

Molecules themselves are made up of **atoms** – sometimes just one, but more frequently several which have been bound together by a chemical reaction. Atoms are the building blocks from which the molecules of a substance are constructed. All atoms are made up of three types of particle – protons, neutrons and electrons (see 1.1).

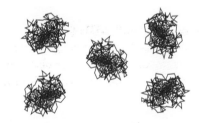

Movement of molecules in a solid,

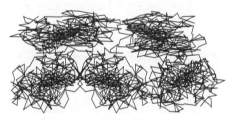

... a liquid,

... and a gas.

In **solids**, each molecule has a permanent linkage with neighbouring molecules due to forces of attraction between them, and so it has a fixed position in the overall arrangement. The molecules can be thought of as 'touching' one another. They are not stationary – each vibrates about a central position as a result of the intermolecular forces. The higher the temperature, the more vigorous the vibration. As a solid gets hotter, a point is reached where the linkages between neighbouring molecules are broken, and the solid becomes a liquid.

In **liquids**, the molecules are still 'touching' their neighbours, but the linkage between them is no longer permanent. The forces of attraction are no longer large enough to keep each individual molecule in a fixed place in the overall arrangement. This is why liquids flow to take up the shape of their container. As in a solid, the molecules vibrate, and the higher the temperature, the more vigorously they vibrate. As a liquid gets hotter, a point is reached where the individual molecules break away from one another to form a gas.

In **gases**, the individual molecules are moving about randomly, and no longer 'touching' their neighbours. The only time one molecule affects another is if they collide or move very close to one another. The molecules move with different speeds, and when they collide their individual speeds will change. The *average* speed of the molecules depends on the temperature of the gas. The hotter the gas, the higher the average speed.

Evidence for moving molecules

Although individual molecules are too small to be seen, what can be seen is the effect of their movement on larger particles. If suitably illuminated smoke particles are observed under a microscope, they appear to shimmer in a jerky, erratic way. The movement of the smoke particles is called **Brownian motion**.

Fast-moving air molecules, which are moving around at random in all directions and with a range of speeds, would be expected to have just such an effect on the smoke particles as they collide with them.

The same effect can be observed in liquids, too. Pollen grains suspended in water are seen to move in a similar way to the smoke particles. Once again, the motion can be explained in terms of collisions.

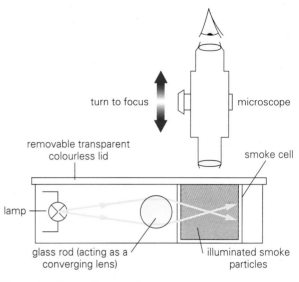

Observing Brownian motion.

Kinetic theory

The theory that all molecules have some kind of motion (and therefore kinetic energy) is called the **kinetic theory of matter**. Kinetic theory is used by scientists to explain a wide range of phenomena, including the expansion of substances as they get hotter (see 5.3), diffusion, gas pressure (see 5.6), evaporation (see 8.3), thermal conduction (see 8.3), and elasticity (see 5.2).

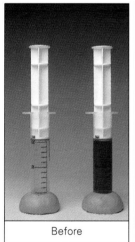
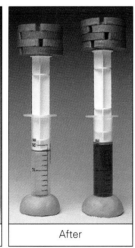

Unlike gases (left, in each photo), liquids are virtually incompressible.

Compressing solids, liquids and gases

Using kinetic theory, it is easy to understand why gases are easily compressed (squashed). As compression takes place, the average space between molecules gets smaller, along with the total volume.

In a solid or a liquid, where the molecules are 'touching', the situation is very different. When the molecules are squeezed together, they start to repel one another. The closer they are squeezed, the greater the repulsive force. So compressing a solid or a liquid by even a small amount requires large forces to be applied. Because of this, we think of solids and liquids as being virtually incompressible.

Solids, such as a sponge, which seem to compress easily are in fact mostly air. As you squeeze a sponge, the trapped air is squeezed out (try putting it under water and then squeezing it).

QUESTIONS

1 What is Brownian motion, and why does it provide evidence for moving molecules?

2 Explain why gases are so much easier to compress then liquids.

3 What are the similarities and differences between the structures of solids, liquids and gases?

5.2 Elasticity

When a pair of equal and opposite forces are applied to a solid, they may change the relative positions of its molecules, and hence its shape. This is what happens when you stretch or compress a spring or a piece of modelling clay.

When the forces are removed, different materials behave in different ways. While a spring returns to its original shape, the clay does not. The more **elastic** a material, the greater its ability to return to its original shape, once distorting forces have been removed.

When a spring is about to become overstretched, it is said to have reached its **elastic limit**. Further stretching leads to permanent damage, so that although the spring will still contract once the force has been removed, it won't return to its original shape. Stretched wires behave in a similar way to stretched springs.

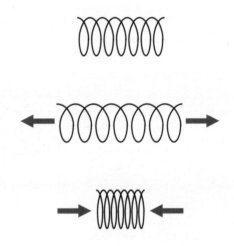

A spring can be stretched or compressed by equal and opposite forces applied to its ends

Hooke's Law

In 1676, Robert Hooke, a contemporary of Newton and fellow member of the Royal Society, announced his discovery that *the extension of a spring or a wire increases in proportion to the load (force) applied*, a rule that has since become known as **Hooke's Law**. Put simply, it means that if you double the load, then the extension will double as well. Like many discoveries in the seventeenth century, this one was announced in the form of an anagram, or puzzle, the idea being to stop unscrupulous rivals from obtaining the information and claiming the discovery as their own!

A spring balance (see 4.1) has even scale markings marked in newtons because each extra unit of force causes the spring to increase in length by exactly the same amount.

Hooke's Law holds only up to the elastic limit. If the load is increased beyond this, a larger extension is produced for the same increase in load, and the spring remains permanently stretched once the load has been removed.

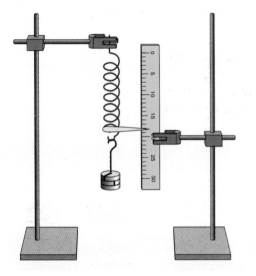

The relationship between a spring's length (and hence extension) and the load applied can be investigated fairly easily.

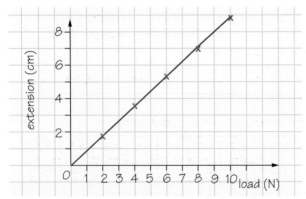

When extension is plotted against load, a straight line passing through the origin shows that Hooke's Law is obeyed.

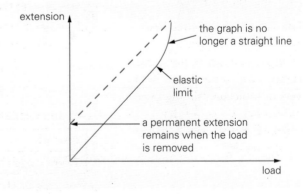

Beyond the elastic limit Hooke's Law is no longer obeyed.

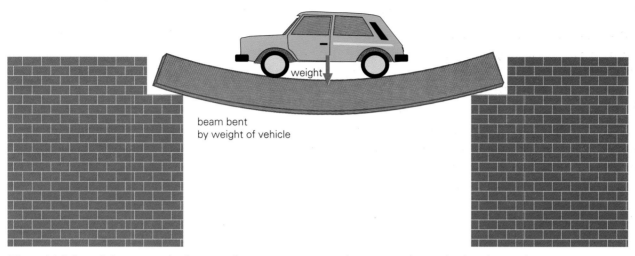

The vehicle's weight causes the beam to become compressed on top and stretched underneath.

Tension and compression
When forces are applied to a solid object which stretch it, it is said to be under **tension**. When the applied forces squeeze or compress the object, it is said to be under **compression**. Structures such as bridges have some parts under tension and other parts under compression. Engineers need to understand the effects when choosing materials and designing structures.

Molecular theory
Elasticity is a result of the forces of attraction and repulsion between the molecules of the material. Normally these forces are balanced. When the molecules in a solid are squeezed together, they repel one another. The closer they get, the greater the repulsive force, and so a greater applied force is needed to squeeze them further together. When the molecules are stretched apart, the force between them becomes one of attraction. The further apart they get (until a certain separation), the greater the attractive force, so a greater applied force is needed to separate them further.

Materials which are stretched beyond their elastic limit undergo a change in molecular structure which involves whole layers of molecules sliding over one another. Once this has happened, a permanent change in shape results.

QUESTIONS

1 A force of 12 N stretches a spring by 6 cm, without exceeding the elastic limit. How far will it stretch if a force of 3 N is applied? What force would make it stretch by 4 cm?

2 Describe, with the help of a diagram, how you would measure the extension of a spring under different loads.

3 A student obtained the following results using a spring:

Load (N)	0.0	0.1	0.2	0.3	0.4	0.5	0.6	0.7	0.8
Extension (cm)	0.0	0.8	1.3	2.0	2.8	3.5	4.1	4.9	5.7

  a Plot a graph of extension against load.
  b If an object hung from the spring caused it to extend by 3.0 cm, what is its weight?

5.3 Density

Most people have heard the riddle 'Which is heavier, a pound of feathers or a pound of gold?' Each has the same mass, and so the same weight. So why do so many people say that the gold is heavier?

Although the feathers and the gold have the same mass, they have very different volumes. People who give the wrong answer have usually taken the volume into account as well, and have given the answer to the question, 'Which is denser?', instead.

The **density** of a material gives you an idea of how much mass there is in a certain volume. The greater the mass in the same volume, the greater the density.

Working out densities

The density of a material can be calculated by dividing its mass by its volume:

$$\text{density (kg/m}^3\text{)} = \frac{\text{mass (kg)}}{\text{volume (m}^3\text{)}}$$

or, in symbols:

$$\rho = \frac{m}{v}$$

The standard unit for density is **kilograms per cubic metre** (kg/m^3), but it is often more convenient in the laboratory to measure it in grams per cubic centimetre (g/cm^3) instead. (One cubic metre is an awful lot of most substances!)

Masses can be easily found using a chemical (top-pan) balance, while volumes can be found in a number of ways, including:
- measuring the dimensions and doing a calculation
- using a measuring cylinder (for small irregular solids as well as for liquids)
- using a displacement can and a measuring cylinder (for larger irregular solids or when more accurate results are required).

Material		Approximate density (g/cm^3)
Metals	Aluminium	2.7
	Copper	8.9
	Gold	19.3
	Iron	7.9
	Mercury	13.6
Woods	Balsa	0.2
	Pine	0.5
	Mahogany	0.8
Other materials	Granite	2.7
	Polythene	0.9
	Pyrex glass	2.2
	Silver sand	2.6
	Water	1.0

The table above shows the densities of some common materials in g/cm^3 at room temperature.

Using less dense materials can reduce weight and cost

The heavier a building, the stronger its foundations need to be and the more expensive they are to build. Some building blocks are much less dense (because they contain large amounts of air) than ordinary bricks. Provided they are structurally suitable, their use will reduce the weight of the building. Volume for volume, they will also be cheaper to transport.

Aluminium alloys are used in aeroplanes because of their combination of lightness (i.e. low density) and strength.

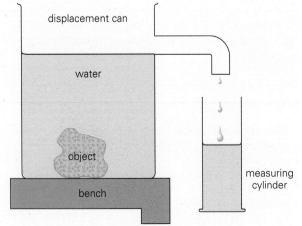

An object will displace its own volume of water.

Floating

Objects that are less dense than water will float in it, while those that are more dense will sink. This can be applied to other liquids: any object that is less dense than the liquid will float, while objects that are more dense will sink. A steel nail will float on mercury (a dense metal which is liquid at room temperature), but sink in water. If two liquids are immiscible (which means that they don't mix together), then the dense one will float on top of the other. This is why the fat floats on top of the gravy!

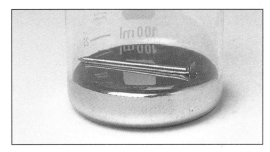

Steel is less dense than mercury, so the nail floats. Beware – mercury is highly poisonous!

Density depends on temperature

As substances are heated, in general they expand due to increased molecular movement, and so take up more space. Consequently, their density will decrease as their temperature rises. In a fluid (where the molecules are not held in a fixed position), any warmer (and therefore less dense) fluid will always rise to the top. This type of movement is known as a **convection current** (see 8.3 and 8.4).

Water, however, behaves differently to other liquids, because as its temperature increases from 0°C to 4°C, it contracts rather than expands. Water at 4°C is denser than water at any other temperature. Ecologically, this is of great significance, because in freezing weather, the water at the bottom of a pond is warmer than the water at the top. As a result, the pond is less likely to freeze all the way to the bottom.

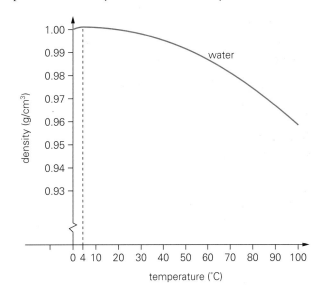

The smoke shows the presence of a convection current due to the heat from the candle flame.

The variation of density of water with temperature.

QUESTIONS

1. Copy and complete this table.

Mass (g)	Volume (cm³)	Density (g/cm³)
10.0	4.0	
25.0		7.1
	10.0	2.7

2. Calculate the density of a piece of wood of mass 450 g, and dimensions 2.0 cm by 2.0 cm by 140 cm.

3. Describe in detail how you would measure the density of a fist-sized lump of rock.

4. Why does a steel ship float in water?

5.4 Pressure

Pressure is a measure of how much force is being exerted on a given area. For a given force, the smaller the area over which it acts, the greater the pressure. Pressure is measured in **pascals** (Pa).

Working out pressure

Pressure can be calculated using the formula:

$$\text{pressure (Pa)} = \frac{\text{force (N)}}{\text{area (m}^2\text{)}}$$

or, in symbols:

$$P = \frac{F}{A}$$

You can see that 1 pascal (Pa) is the same as 1 newton per square metre (N/m^2). For a given area, doubling the force will double the pressure. The pressure can also be doubled by keeping the force the same and halving the area.

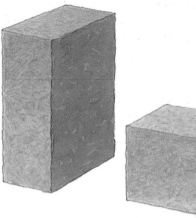

Although both bricks have the same weight, the brick standing on end is exerting twice the pressure, because the face it is standing on has half the area.

How much pressure?

A drawing pin being pushed into a piece of wood	1 000 000 000 Pa
A person standing on one stiletto heel	10 000 000 Pa
Atmospheric pressure	100 000 Pa
A suitcase lying on the floor	1 000 Pa
A coin resting on a table	100 Pa
A sheet of paper resting on a table	1 Pa

Both of these filled bags have the same weight, but the one on the left would be more painful to hold.

That sinking feeling!

Many everyday solids will dent and be permanently damaged or marked if enough pressure is applied to them. Some dent very easily; others are able to withstand much greater pressures.

Sometimes things are deliberately shaped so that they will sink into things more easily – by making the contact areas smaller. Knives, pins and nails are good examples. On the other hand, the foundations of buildings are built wider than the walls, to reduce the pressure on the ground. Similarly, skis and snowshoes are designed with a large contact area, which helps reduce the amount of sinking.

Snowshoes stop you sinking!

Stiletto heels can damage floors!

Pressure in liquids

When a pressure-measuring device is lowered into a liquid, the deeper it goes, the greater the reading. If, however, it is stopped at one particular point, and then rotated, the reading stays the same, no matter which way it is facing. From this it follows that:

- the pressure in a liquid increases with depth
- at any point in a liquid, the pressure acts equally in all directions.

The unit of pressure is named after Blaise Pascal (1623–1662), a French mathematician and physicist.

Consider what would happen if the pressure in one particular direction was greater than in another – there would be a resultant force in that direction, which would make the liquid move in that direction too!

The same principles apply to gases as well as liquids – atmospheric pressure, for example, decreases with increasing height above the Earth's surface.

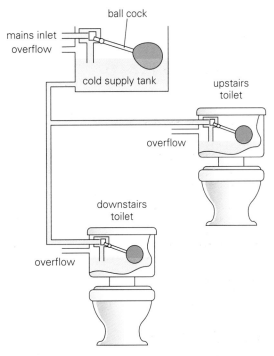

The deeper you dive the greater the water pressure, and too much pressure can damage your ear-drums.

The cistern of a toilet downstairs fills more quickly than one upstairs because of the greater difference in height between it and the tank in the loft.

QUESTIONS

1. Explain why the left-hand bag shown opposite would be more painful to carry than the other bag.

2a. Make a list of things whose contact area has been deliberately increased to prevent them from sinking into surfaces.
  b. Make a second list of things whose contact area has been deliberately made as small as possible to make them sink into surfaces more easily.

3. Calculate the pressure exerted on a table by a lamp whose weight is 12 N and whose square base measures 0.1 m by 0.1 m.

4. Calculate the pressure exerted on a floor by an elephant of weight 40 000 N and combined foot area of 0.4 m^2. Which would cause more damage to the surface of a wooden floor, a person standing on one stiletto heel or an elephant?

5. Why are dams thickest at the bottom?

5.5 Hydraulic systems

The digger and the crane are both **force multipliers** – machines designed to enlarge the size of the applied force – but they work in different ways. A crane uses pulleys to magnify the forces. The digger uses **hydraulics**.

These machines are force multipliers.

What is a hydraulic system?

In a hydraulic system, a liquid is used to transmit a force to where it is needed. A force (the effort) is applied to a liquid using the master cylinder and piston. This puts the liquid under pressure. Liquids are virtually incompressible, so the pressure is transmitted instantaneously and evenly throughout the liquid. This causes a force to be applied at the slave cylinder on the slave piston. The size of the force depends on the relative cross-sectional areas of the pistons. It does not depend on the shape or angle of the interconnecting pipework.

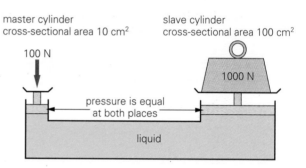

A simple hydraulic system. For each centimetre the 100 N moves downwards, the 1000 N weight moves upward by just 0.1 cm.

How much is the applied force enlarged?

Since the pressure is the same everywhere within the system, the pressure on both the pistons must be the same. You have already seen that:

$$\text{pressure (Pa)} = \frac{\text{force (N)}}{\text{area (m}^2\text{)}}$$

Therefore:

$$\frac{\text{force applied to master cylinder (N)}}{\text{cross-sectional area of master cylinder (m}^2\text{)}} = \frac{\text{force applied to slave cylinder (N)}}{\text{cross-sectional area of slave cylinder (m}^2\text{)}}$$

This can be rearranged to give:

$$\text{force applied to slave cylinder (N)} = \text{force applied to master cylinder (N)} \times \frac{\text{cross-sectional area of slave cylinder (m}^2\text{)}}{\text{cross-sectional area of master cylinder (m}^2\text{)}}$$

Free work?

Although the system enlarges a force, it cannot enlarge the amount of **work** put in. Doing work increases the energy of a system – work, like energy (see 8.8), is measured in joules and is calculated by using the formula:

$$\text{work done (J)} = \text{force (N)} \times \text{distance moved in direction of force (m)}$$

If the slave cylinder has 10 times the area of the master cylinder, then the force will be multiplied by a factor of 10. But the distance moved by the master piston will be 10 times greater than the distance moved by the slave piston. This means that the work done on the master cylinder by the effort is the same as the work done by the slave cylinder on the load. You can never get more work out of a machine than you put into it!

Advantages of hydraulic systems
- They can be used as force magnifiers.
- They can be used to apply forces in any chosen direction – the connecting pipes can be made flexible to make them even more versatile.
- They can apply a force to more than one slave cylinder at a time.

Hydraulic braking systems

When the brake pedal is pressed, a force is applied to the piston of the master cylinder. The four slave cylinders apply a force to the brake pads to stop each of the wheels turning. All four slave pistons have the same cross-sectional area, so all apply the same braking force. If different forces were applied to each of the wheels, the car would be very difficult to steer.

If air gets into the hydraulic system, the brakes feel 'spongy' and don't work properly. This is because when the master piston is pushed, the air becomes compressed. Brake systems are designed so that any air which enters the system when the hydraulic liquid is being changed can be easily removed or 'bled' from the system.

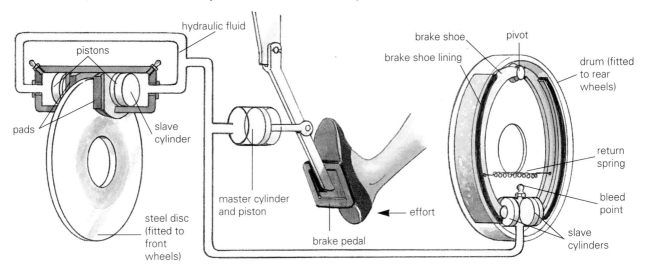

Cars have hydraulic braking systems.

QUESTIONS

1. How many machines can you think of which use hydraulics as force magnifiers? Make a list.

2. If a hydraulic system magnifies a force by a factor of 20, and the cross-sectional area of the master cylinder is $2\,cm^2$, what is the cross-sectional area of the slave piston?

3. Work out the force which must be applied to the master piston of a hydraulic system if the slave piston is to raise a load of 100 000 N, and the cross-sectional area of the slave cylinder is 250 times greater than that of the master cylinder.

4. By referring to this diagram of a hydraulic jack, describe in your own words how it works.

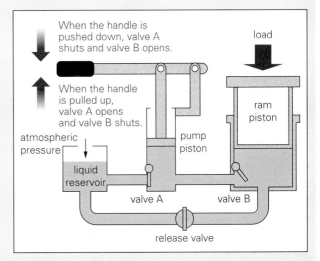

Hydraulic jacks are designed to raise heavy loads. The system of valves allows them to operate safely.

5.6 Pressure in gases

Why gases exert a pressure

The molecules in a gas move around freely and rapidly in all directions, colliding with the walls of the container. Each time a molecule strikes a wall, it does so with a certain force. Each second, any given part of the container's surface will be hit many millions of times. The combined force of all these collisions gives rise to the pressure within the container.

The air in an empty drinks bottle consists of billions of billions of molecules, each travelling with an average speed of around 500 m/s.

Pressure increases with temperature

If a sealed container of gas is heated, then the pressure of the gas will increase. This can be explained by considering what happens to the molecules. The hotter the gas, the greater the average speed of the molecules. This means that in a period of one second, not only will more molecules hit an area of the container wall, but they will also hit it harder. In other words, they will be hitting the surface with a greater force than they were before. As a result, the pressure increases.

Pressure depends on volume too

In the seventeenth century, the Irish scientist Robert Boyle investigated how changing the volume of a fixed amount of gas at a constant temperature, affected its pressure. He found that halving the volume, doubled the pressure.

He showed that:

initial pressure × initial volume = final pressure × final volume

or, in symbols: $p_1 \times V_1 = p_2 \times V_2$ or $p_1 V_1 = p_2 V_2$

This rule is known as Boyle's Law.

As the volume decreases, the gas pressure will rise; and as the volume increases, the pressure will fall. The total number of gas molecules doesn't change, but reducing the volume of the container will squeeze them closer together. The average speed of the molecules remains the same (because the temperature is constant), but because any given space now contains more molecules, more will collide with the container's surface each second and so the pressure will be greater. Similarly, increasing the volume of the container results in fewer collisions each second and so reduces the pressure. The pressure does not depend on the *shape* of the container, only the volume.

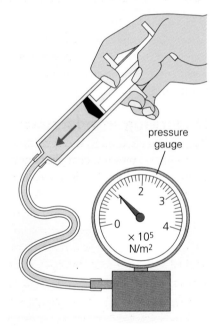

Boyle's Law can be investigated with this apparatus.

The atmosphere can exert large forces

Rubber 'suction pads' (like those on some soap and towel holders) stick to smooth surfaces, because of the pressure exerted by the atmosphere. As you press them into position, the rubber suckers flex and some of the air is forced out from beneath them. When you let go, the rubber springs back, reducing the pressure of the air beneath the pad. The difference in pressure between the air beneath the pad and the air outside provides the force that holds the pad firmly in place.

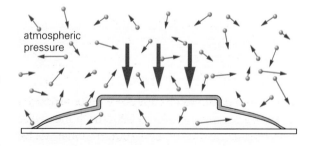

Rubber suction pads stick to a smooth surface because of atmospheric pressure.

Otto von Guericke's experiment.

In the seventeenth century in Magdeburg in Germany, Otto von Guericke demonstrated just how large the forces exerted by the atmosphere are. He held two metal hemispheres together to make a sphere, and then pumped out the enclosed air. Neither 50 men, nor two teams of horses were able to pull the hemispheres apart!

Measuring atmospheric pressure

Barometers are used to measure atmospheric pressure. Atmospheric pressure changes continuously with weather conditions, but is normally between 94 000 and 106 000 Pa at sea level. Meteorologists use a unit of pressure called the bar. One bar is equal to 100 000 Pa. Aneroid barometers (the type often found in people's hallways), consist of a sealed springy metal box, which has had some of the air pumped out of it. As the atmospheric pressure rises, the box is crushed more, and gets slightly smaller. When the pressure falls, the reverse occurs. A system of levers allows the small movements of the box to be magnified and shown on a scale.

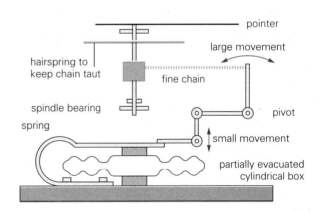

The workings of an aneroid barometer.

QUESTIONS

1. If the temperature is kept constant, by what factor will the volume change when the pressure of a fixed amount of gas in a syringe is **a** doubled, **b** halved, **c** increased by a factor of ten?

2. Gas at a pressure of 105 000 Pa is contained in a sealed canister of volume 500 cm^3. If the canister is crushed to a volume of 100 cm^3, what will be the pressure inside it, assuming the temperature doesn't change?

3. Explain why the pressure of a fixed amount of gas decreases when its volume is increased, at constant temperature.

4. Why doesn't the pressure inside a balloon decrease as it is blown up?

5. Why won't a suction pad stick to a rough surface?

SECTION B: QUESTIONS

For these questions take the Earth's gravitational field strength g as 10 N/kg.

1. Explain the following:
   a. If a lorry and a military tank both have the same mass, the lorry is more likely to sink in soft ground than the tank.
   b. A hole at the bottom of a ship is more dangerous than one just below the water's surface.

2.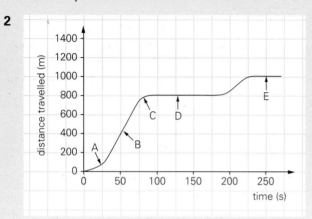

   The graph above shows distance plotted against time for a short car journey to a school. At which of the points A – E is the car:
   a. stationary?
   b. travelling at a steady speed?
   c. speeding up?
   d. slowing down?
   e. travelling with the greatest speed?
   Describe how the speed of the car varied during the course of the trip.

3. The graph below shows a velocity–time graph for a car travelling in a straight line.
   a. What was the maximum speed of the car?
   b. How long did the journey take?
   c. What was the acceleration of the car during the three stages of its journey?
   d. For how long did the car travel at a steady speed?
   e. What is the distance it travelled during this time?
   f. What was the total distance travelled?

   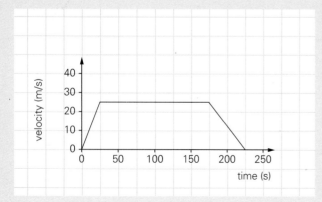

4. An aircraft can produce a maximum thrust of 500 000 N. It has a mass of 150 000 kg.
   a. What is its weight?
   b. Draw a diagram showing the size and direction of the forces which are acting on the aircraft in flight, assuming it is flying in a straight line at a steady speed, at a steady height, with its engines producing their maximum thrust.

5. The graph shows how the drag on a motor boat increases with speed.

   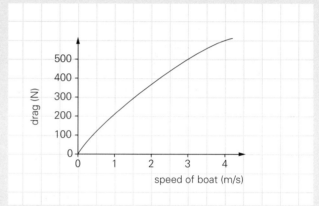

   a. What is the size of the drag when the boat is travelling at 3 m/s?
   b. What size driving force would be needed to make the boat travel at 3 m/s?
   c. How far will the boat travel in 30 s, at a steady 3 m/s?
   d. If the boat was stationary and its engine fell off, how might the velocity of the engine change as it fell to the sea bed several hundred metres below? Explain why it would change in this way.

6a. Explain why it is easier to loosen a nut if a spanner with a longer handle is used.
  b. Why is it usually not a good idea to increase the length of a spanner as shown when tightening a nut?

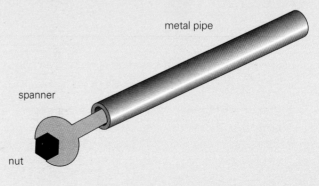

7. The pressure exerted by a chair on the floor is 50 000 Pa. The contact area of each of the four legs is 5 cm².
   a. Calculate the weight of the chair. (Remember to convert the contact area in cm² to one in m².)
   b. What pressure would be exerted if a person whose weight was 700 N sat in the chair with his legs off the ground?
   c. If the average density of the chair is 1.2 g/cm³, what is its volume?
   d. If the chair was thrown into a pond, would it float or sink? Explain your answer. (Density of water = 1.0 g/cm³.)

8a. If the atmospheric pressure is 100 000 Pa, what force does the atmosphere exert on the side of a wall measuring 10 m by 2 m?
   b. Why doesn't the wall topple over?

9. A force of 4000 N is applied to the master cylinder of a hydraulic press whose cross-sectional area is 4 cm². What force will be exerted by the slave cylinder if it has an area of 100 cm²?

10. The density of air is 1.3 kg/m³. What is the mass of air contained in:
   a. a room measuring 3 m × 2.5 m × 4 m?
   b. an 'empty' 1 litre drink bottle?
   c. If the atmospheric pressure is 100 000 Pa, what would the pressure and the density of the air inside the bottle be if it was crushed to two-thirds of its volume, **i** with the cap on and **ii** with the cap off?

11. In an experiment with a spring, the following results were obtained.

Load (N)	0.0	1.0	2.0	3.0	4.0	5.0	6.0
Length of spring (cm)	5.0	5.8	6.5	7.3	8.1	9.4	9.5

Plot a graph of extension against load.
   a. Which set of measurements has probably been incorrectly made or recorded?
   b. What weight objects would cause the spring to increase in length to: **i** 6.8 cm and **ii** 6.5 cm?

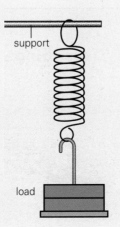

12. Identical balls, each of mass 0.1 kg, were thrown upwards at an angle of 45°, with identical initial velocities. However, while one was thrown on the Earth, the other was thrown on the Moon. The gravitational field strength on the Moon is approximately 1.7 N/kg.
   a. What would be the difference in the balls' weights?
   b. On the same set of axes, sketch a pair of graphs to show how the height of each ball might vary with time.
   c. On a second set of axes, sketch a pair of graphs to show how the horizontal distance travelled by each ball might vary with time.
   d. Explain any similarities or differences between the graphs you have drawn.

13. Explain why smoke alarms are placed on the ceiling, rather than at floor level.

14. A lump of quartz containing veins of gold has a mass of 588 g and a volume of 100 cm³. If the density of quartz is 2.6 g/cm³, and the density of gold is 19 g/cm³, what is the mass of gold contained in the quartz?

15. Aerosol cans normally carry the written warning 'Pressurised container: protect from sunlight'. Use kinetic theory to explain what might happen if the warning was ignored.

CHAPTER 6: CHARACTERISTICS OF WAVES

6.1 What are waves?

The people in the photograph are enjoying playing in the waves. We are all familiar with water waves, but waves affect our lives in many more ways than perhaps we might first imagine. Light travels as a wave and so does sound; most of us make use of radio waves every day when listening to the radio or watching the television or making a telephone call. To understand how waves behave we must first know how to describe them.

Different types of wave

A long 'slinky' spring can be used as a good model for wave behaviour. You can make two different types of wave travel along the spring.

In **a**, the wave is made by moving the end of the spring from side to side, at right angles to the spring. This produces a **transverse wave** – transverse pulses move along the spring. Water waves are an example of transverse waves.

In **b**, the end of the spring is being pushed and pulled, making each coil in the spring vibrate back and forth, parallel to the length of the spring. The result of this is that the coils of the spring bunch together in places and pull apart in other places, and the whole pattern moves along the spring. This sort of wave is a **longitudinal wave**. Sound waves are an example of longitudinal waves.

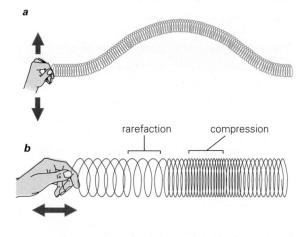

Two different waves travelling along a slinky spring: **a** a transverse wave; **b** a longitudinal wave.

In both types of wave each coil of the spring simply moves back and forth about one point; this movement is called **oscillation**. In a transverse wave the coils oscillate at 90° to the direction of the wave movement. In a longitudinal wave the oscillation is in the same direction as the wave movement. In neither case is there any movement of the spring as a whole. So what *is* travelling through the spring? As pulses arrive at the other end of the spring you can feel them with your hand. The wave is carrying *energy* along the spring without bringing the coils of the spring with it.

In the case of water waves, or sound waves in air, it is the particles of water, or air, which oscillate. Energy is transferred but there is no overall movement of matter: *so a wave is a transfer of energy without a transfer of matter.*

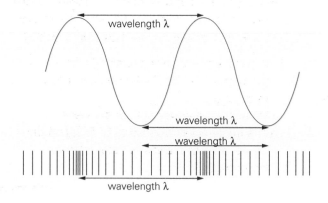

Describing waves

Waves are created by a disturbance or a vibration. In the case of the slinky spring demonstration, the disturbance is the person shaking the end of the spring. If the disturbance is regular, as the waves travel along the spring you can see a repeated pattern. The distance between a point and the next point along the spring where the pattern repeats is called the **wavelength**. Wavelength is measured in metres and the Greek letter λ (lambda) is used to represent it in diagrams and equations.

The number of complete waves produced each second is called the **frequency**, f. The unit of frequency is **hertz** (Hz). A vibration of say 20 vibrations or oscillations per second creates a wave with a frequency of 20 Hz. This means the wave is being produced at a rate of 20 wavelengths every second.

Higher frequencies are measured in **kilohertz** (kHz) or **megahertz** (MHz).

1 kHz = 1000 Hz

1 MHz = 10^6 Hz

The distance from the central point of an oscillation to the point of maximum displacement from that central point is called the **amplitude**. The amplitude of a longitudinal wave on the slinky spring is a measure of the difference between the normal density of the coils and their maximum compression.

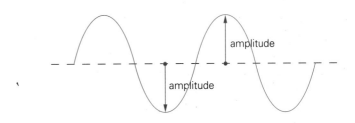

The amplitude of a transverse wave is simple to visualise.

Two longitudinal waves of the same wavelength but different amplitudes.

The speed of a wave

Waves take time to travel. The speed of a wave is given by:

$$\text{speed (m/s)} = \frac{\text{distance the wave travels (m)}}{\text{time taken (s)}}$$

If a wave is travelling at 5 m/s with a frequency of 5 Hz then in one second the first wavelength would have travelled 5 m, and five wavelengths would have been produced altogether. The wavelength of the wave would therefore be 1 m. If the wave's frequency was 10 Hz then its wavelength would have to be 0.5 m to fit ten waves into the same 5 m space. Wavelength therefore depends upon speed and frequency:

speed (m/s) = frequency (Hz) × wavelength (m)

or, in symbols:

$v = f \times \lambda$

For example, if water waves are being produced with a frequency of 15 Hz and their wavelength is 3 cm, what is the speed of the wave?

$v = f \times \lambda$

$= 15 \times 0.03 = 0.45$ m/s

QUESTIONS

1a What is the difference between a transverse wave and a longitudinal wave?
b Write down two things that transverse and longitudinal waves have in common.

2a A sound wave has a frequency of 100 Hz and a wavelength of 3.30 m. How fast does the wave travel?
b A radio wave has a frequency of 198 kHz and a wavelength of 1515 m. How fast does it travel? (Give your answer to two significant figures.)

3 A water wave is being produced with a frequency of 25 Hz and travels at a speed of 5 m/s.
a Work out the wavelength of the wave.
b The frequency is reduced to 12.5 Hz and the speed remains the same. What is the new wavelength?

6.2 Looking at waves in a ripple tank

You can use a ripple tank to study the behaviour of water waves. A vibrating bar produces transverse waves across the surface of the water. The number of times this bar moves up and down every second is the frequency of the wave. The time it takes the bar to make one complete movement, all the way up and back down again, is called the **time period** or simply **period** of oscillation, measured in seconds. The relationship between the frequency and the period of a wave is:

frequency (Hz) = $\dfrac{1}{\text{period (s)}}$

or, in symbols:

$f = \dfrac{1}{T}$

The wavelength of water waves is the distance between consecutive 'crests' (or 'troughs') of the ripples. As the speed at which the bar vibrates is increased (increasing the frequency), the wavelength can be seen to become shorter. The speed of the wave stays constant.

Different wave effects can be demonstrated using a ripple tank.

The top ripple tank is producing waves at a lower frequency than the one on the bottom.

Reflection

As you know, waves carry energy. A wave will keep on going until all the energy is dissipated (lost to the surroundings). Think of water waves on a beach. When a wave arrives at the shore, some of the energy is used to make the noise of the wave breaking on the shore, some of it is used to move the sand and pebbles around and some of it is turned into heat. If the wave meets an obstruction that will not absorb the energy of the wave, then the wave will be **reflected**.

You can see this happening if you put a rigid barrier in a ripple tank. As the waves strike it, only a tiny proportion of their energy is absorbed – most is retained, so the waves are seen to be reflected. There is no change in speed or wavelength of the waves. The angle the reflected wave makes with the barrier is equal to the angle the incident wave makes with the barrier.

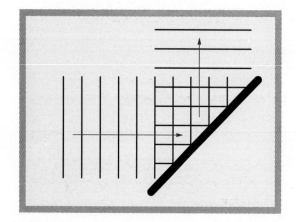

Ripples are reflected at a straight surface.

Refraction

The speed of a water wave depends on the depth of the water it is travelling in – a wave will travel more quickly in deep water than in shallow water. Consequently, if the depth of water decreases the wave will slow down. This is why sea waves 'break' as they approach the shore. The bottom of the wave slows down in the shallow water, but the top continues at the same speed, so the wave breaks.

The depth of water in a ripple tank can be changed by putting a thick piece of transparent plastic in the tank. The effect of this is shown in the diagrams.

Waves break because of a difference in speed between the top and the bottom.

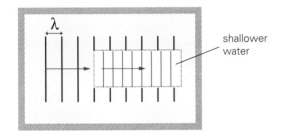

The wavelength of the water waves becomes shorter in the shallow water.

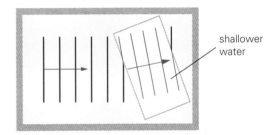

When the decrease in depth occurs at an angle to the incident waves, the waves change direction.

The water waves slow down as they reach the shallower region. Since speed = frequency × wavelength, as the speed decreases the wavelength must also decrease because the frequency has not changed.

If the boundary between the deeper and the shallower water makes an angle with the incident waves, the direction of travel of the waves will change. This effect is called **refraction**. To understand why this happens, consider a straight 'wavefront' (line of crests, say) approaching the boundary. All parts of the wavefront do not reach the oblique boundary at the same time, so do not slow down at the same time. This causes the wavefront to 'swing around'. The same happens to all the incident wavefronts, so they continue in the shallower region in a different direction.

All types of wave (except 'one-dimensional' waves along a spring or a rope) can be refracted. Sound waves travel at different speeds in different materials and so will be refracted. Light waves are refracted when they enter and leave transparent materials such as glass or Perspex.

QUESTIONS

1. The vibrating bar in a ripple tank makes one complete movement all the way up and back down again every 0.05 s. What is the frequency of the wave produced?

2. Why won't a water wave be reflected well from an obstacle made of a soft material like foam rubber?

3a. What does the word *refraction* mean?
 b. Why does a change in wave speed sometimes cause refraction?

6.3 Electromagnetic waves

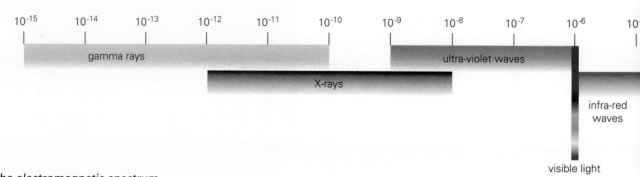

The electromagnetic spectrum.

The electromagnetic spectrum

A very important group of waves is the spectrum of electromagnetic waves – the **electromagnetic spectrum**. It includes visible light, radio waves, microwaves, ultra-violet and infra-red waves, X-rays and gamma rays.

Like any other wave, electromagnetic waves can be reflected and refracted, but they are unique in the way they travel. *Electromagnetic waves are the only waves that can travel through a vacuum.* They are transversely oscillating electric and magnetic fields that don't require the presence of particles.

Electromagnetic waves travel very fast indeed, with a speed of 3×10^8 (300 000 000) m/s in a vacuum. Concorde would take 16 hours to travel all the way around the world, travelling at twice the speed of sound. An electromagnetic wave could travel the same distance in less than 0.2 seconds!

The difference between the differently named waves that make up the electromagnetic spectrum is in the wavelength and hence frequency of the waves. The illustration shows the complete range.

Electromagnetic waves of different wavelength ranges are produced in different ways and have different properties. They are reflected, transmitted and absorbed differently by different materials. They all transfer energy, and cause an increase in temperature when absorbed. They are often collectively called **electromagnetic radiation**.

Because of their different effects, we make use of the different wavelength ranges in different ways.

Radio waves

Radio waves are created by the oscillation of electrons in a conductor. They are used to transmit radio and TV signals between different points on the Earth's surface. When radio waves reach a receiving aerial, some of their energy is absorbed. This energy makes the electrons in the metal aerial oscillate at the same frequency as the radio wave. These oscillating electrons make up an alternating current (see 2.1) which is then amplified by the receiving equipment.

Radio waves have a very wide range of wavelengths, from a few kilometres to a few centimetres. Long wavelengths can be used to transmit signals over very long distances. Like all electromagnetic waves, radio waves travel in straight lines. However, long wavelengths are reflected by the electrically charged layer in the Earth's upper atmosphere. This means that they can be sent all around the world despite the Earth's curvature.

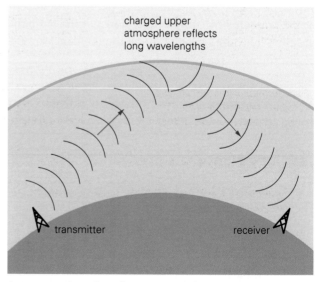

Long-wavelength radio waves can be transmitted around the world by using the upper atmosphere as a reflector.

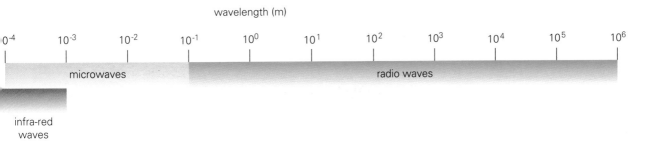

Microwaves

Microwaves are of course used in microwave ovens. These ovens heat food very quickly. They do this by transmitting microwaves with a frequency which is the same as the natural frequency of vibration of a water molecule. When these microwaves are directed through food, the frequency match means that all of their energy is absorbed by the water in the food. The water molecules vibrate very vigorously because of this extra energy and consequently the food becomes hot. (The faster that molecules move, the higher the temperature – see 8.3.)

Living tissue, which is mainly made up of water, can be seriously heat-damaged very quickly by microwaves of the frequency used in microwave ovens. This is why the ovens need to be carefully screened to stop any microwave radiation escaping.

The dishes on this tower are microwave links.

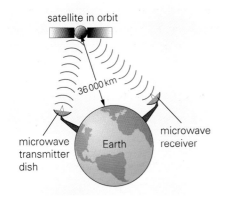

Microwaves are used for long-distance telephone and TV transmissions via satellites.

A very important use of microwaves is for communication. Information, such as speech or data, can be converted to a varying electrical signal and then transmitted from an aerial as microwaves (basically short-wavelength radio waves). Large quantities of information are sent up and down the country between telecommunications towers. Mobile telephones also use this principle.

Microwaves are particularly useful for communicating via satellites. Communications satellites are in geostationary orbit (see 11.9) at a height of about 36 000 km. The short wavelength of a microwave transmission means that it can pass through the Earth's upper atmosphere without being reflected in the way that longer-wavelength radio transmissions are. Microwaves also spread out very little when focused using a reflecting dish. Both of these properties make microwaves ideal for communicating with distant satellites.

QUESTIONS

1. How do electromagnetic waves differ from all other waves?

2. Write out this list of different electromagnetic waves in order of increasing wavelength: gamma rays, radio waves, infra-red, ultra-violet, microwaves, visible light, X-rays.

3. Explain why long-wavelength radio waves cannot be used to communicate via satellites.

4a. What are the hazards associated with microwaves?
 b. Why are microwaves used for communication links no great danger to birds?

6.4 Infra-red and ultra-violet

Infra-red waves

The most familiar use for infra-red radiation is in heating, for example, by an electric grill or an electric bar fire. Any heating element that glows visibly red or orange when it is switched on is radiating mostly infra-red waves.

The energy of infra-red waves is absorbed by your skin, making you feel warm. Excessive exposure can lead to skin burns.

Everything emits (gives out) some infra-red radiation. The wavelength range emitted depends on the temperature of the object. Cameras that are sensitive to infra-red rather than visible wavelengths can therefore be used at night, or in other very dark and difficult conditions, to see anything that is at a higher temperature than its surroundings.

Video cameras that are sensitive to infra-red light can be used to see at night.

Satellites in low orbits can be used to scan the Earth (see 11.8). When equipped with infra-red cameras they provide images showing land use, for example different types of crops, deforestation, conurbations.

Specialist infra-red detectors have been used, very successfully, to find survivors of earthquakes trapped underneath piles of rubble up to two weeks after the earthquake took place.

Perhaps the most common use for infra-red waves these days is in remote controls for televisions, videos and hi-fis. An infra-red light-emitting diode (LED) is mounted at the front of a remote-control handset, and an infra-red sensitive receiver is mounted on the front of the equipment. When a button is pressed on the handset the LED emits a particular sequence of infra-red pulses. The receiver decodes this and carries out the instruction.

A remote control sends a coded series of infra-red pulses to the television.

Ultra-violet waves

Some of the energy radiated by the Sun is in the form of ultra-violet waves, which have wavelengths shorter than visible light. Ultra-violet waves can tan the skin, but they also penetrate the top layer of skin and can cause the living tissue underneath to burn if you stay out in the sun too long. Very high-energy (short-wavelength) ultra-violet radiation can cause serious damage to the cells of living tissue, possibly leading to skin cancer. The darker the colour of the skin, the more ultra-violet energy is absorbed by the outer layers and so less penetrates to harm the deeper tissue. Eyes can also be seriously damaged by over-exposure to ultra-violet radiation.

Be careful if you sun-bathe. An increasing number of people are suffering from skin cancer through too much exposure to ultra-violet from the Sun.

Fortunately, most of the hazardous shorter wavelengths are filtered out by a layer of gas high in the upper atmosphere called the ozone layer. Ozone is a molecule made up of three atoms of oxygen (O_3). Many scientists believe that chemicals called chloro-fluoro-carbons, which are traditionally used in refrigerators and aerosols, are causing 'holes' in the ozone layer.

People who like to have a sun tan may sometimes use a 'sunbed'. Ultra-violet radiation is created by using a series of glass tubes filled with low-pressure gas. When a voltage is applied across the ends of a tube, the gas in it conducts electricity and, as it does so, it emits low-energy ultra-violet radiation which gives a good tan relatively safely.

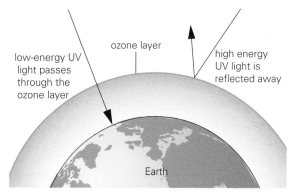

The ozone layer around the Earth's atmosphere gives us some protection from harmful ultra-violet radiation.

Fluorescent strip lights use a very similar principle to sunbed lamps. The difference is that the outside of the tube is coated with a special **fluorescent** material. When the ultra-violet radiation hits the fluorescent coating, the energy is absorbed and then re-emitted at a visible wavelength. This effect of fluorescence is made use of in invisible security markings on valuable items – when exposed to ultra-violet radiation the marks fluoresce and can be read.

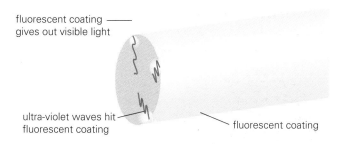

Ultra-violet radiation created inside the tube of a strip light is changed into visible light by the fluorescent coating.

QUESTIONS

1. Explain why ultra-violet waves can pass through the top surface of skin but infra-red waves cannot.

2a. What is the ozone layer?
 b. Why should we be worried if the ozone layer is damaged?

3. Explain how cameras sensitive to infra-red waves can be used to 'see' at night.

6.5 X-rays and gamma rays

X-rays

Electromagnetic radiation with wavelengths in the region of 10^{-8} to 10^{-12} m are called X-rays. They were accidentally discovered by a German physicist called Wilhelm Röntgen in 1895, while he was experimenting with gas discharge tubes (like the ones used for fluorescent lighting).

Röntgen found that X-rays made photographic film blacken. They are also able to pass through soft living tissue easily, but are almost stopped by bone. These properties mean that X-rays can be used to produce shadow photographs of the inside of our bodies; their use revolutionised medicine at the beginning of this century.

X-rays are stopped completely by metals. This makes them particularly useful for security scanning at airports. All luggage is X-rayed before it is allowed on an aeroplane, to see if any guns or bombs are present. In industry, faults in metal machinery can be found by examining a radiograph.

Wilhelm Röntgen, 1845–1923

X-rays are produced when highly accelerated electrons are fired at a metal target. Commercial X-ray machines are evacuated tubes with a target made of a metal such as tungsten which is able to withstand the very high temperatures generated. The tube is shielded with lead, with a small 'window' to allow a beam of X-rays out. The wavelength range of the X-rays produced can be controlled by adjusting the speed of the incident electrons.

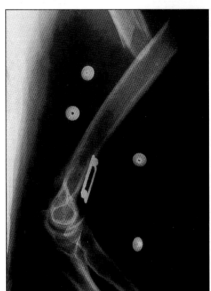
X-ray shadow photographs or 'radiographs' are particularly useful for diagnosing broken bones.

X-rays have to be used in carefully controlled amounts as they can damage or even kill cells by ionisation (see 10.2). Over-exposure to X-rays can cause permanent damage, including cancer.

Short-wavelength X-rays can be used to treat cancer by killing the cancerous cells.

Gamma rays

Unstable nuclei of radioactive elements emit large amounts of energy in the form of very short-wavelength electromagnetic waves called gamma rays. Gamma rays are very harmful to living things. Their high energy and hence high ionising ability can kill cells instantly, or cause them to become cancerous at lower levels of exposure.

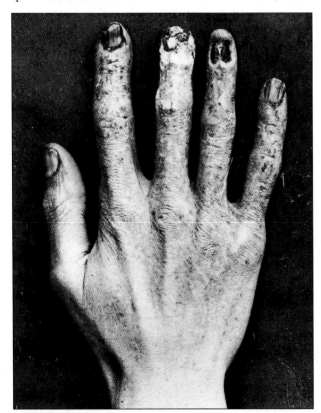
This doctor's hand has become severely damaged from using X-rays with no protection.

Gamma radiation is often used in hospitals for sterilising surgical instruments, as single-celled bacteria are killed instantly upon exposure.

It can be used similarly to kill harmful bacteria in food and so extend the shelf-life of the food. This procedure is rather controversial, as some people suspect that bacteria-infected food could be sterilised and sold even though it is still unfit for consumption.

A very important application of gamma radiation is in the treatment of cancer. Cancerous cells can be killed using gamma-emitting radioactive sources. Gamma rays are more effective than X-rays because of their higher energy. Careful focusing of the gamma rays avoids damage to surrounding healthy cells.

Gamma emitters are also used in medicine as 'tracers' inside the body, enabling, for example, tumours to be detected (see 10.6).

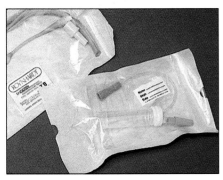

Gamma radiation is used for sterilising medical equipment.

Summary of the hazards and uses of electromagnetic waves

Type of wave and wavelength	Uses	Hazards
radio waves 10^6 to 10^{-1} m	broadcasting, communications, radar	none proven
microwaves 10^{-1} to 10^{-4} m	communications including satellite transmission, cooking	some wavelengths cause rapid heating of skin tissue leading to severe burns
infra-red waves 10^{-3} to 10^{-6} m	electric bar heaters, grills, night photography, remote-control handsets	skin burns
visible light 10^{-6} m	seeing, photography, communication	none
ultra-violet waves 10^{-6} to 10^{-9} m	sunbeds, fluorescent strip lights, security marking	skin burns; the shorter wavelengths cause surface cell damage leading to skin cancer
X-rays 10^{-8} to 10^{-12} m	photographing inside the body, treating cancer, airport security, identifying faults in machinery	severe cell damage leading to cancer or destruction of cells
gamma rays less than 10^{-10} m	treating cancer, sterilising medical equipment, sterilising food	severe cell damage leading to cancer or destruction of cells

QUESTIONS

1. Why are X-rays useful for diagnosing broken bones?

2. Why do radiographers stand behind a lead glass screen when they operate an X-ray machine?

3. Using the wavelength ranges in the table above, work out the frequency range for each group of electromagnetic waves and enter it into a table. (Use the formula speed = frequency × wavelength; speed of electromagnetic waves = 3×10^8 m/s.)

4. Why is food treated with gamma radiation required to be labelled 'irradiated'?

6.6 Light is a wave

Visible light is part of the electromagnetic spectrum, so it has the following properties:

- it is a transverse electromagnetic wave
- it travels at 3×10^8 m/s
- it travels in straight lines
- it can travel through a vacuum.

In common with all other types of wave, light can be reflected and refracted.

Reflection of light

When a wave meets an obstruction it is reflected. You have seen that this happens with water waves and radio waves. Light is reflected in just the same way; in fact this is how you see things – light falls upon an object and is reflected into your eyes.

Light reflected from an object is usually scattered in all directions. The person can see the object because some of the light enters her eyes.

The appearance of an object depends in part upon how smooth it is. The more highly polished or 'shiny' an object is, the more efficiently it will reflect light. If a surface is very shiny indeed, you see an **image** of yourself in the surface. Mirrors are extremely good reflectors because they are very smooth and shiny.

Light from a point on an object arrives at a distant surface travelling in uniform, parallel lines or **rays**. If the light meets a rough reflecting surface then it will be scattered in all directions. This is called **diffuse reflection**.

However, if the surface is smooth then the light will be reflected uniformly, preserving the information the light waves were carrying and hence producing an image of the object. This is called **regular reflection**.

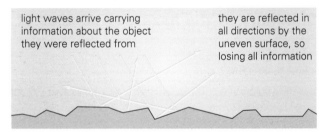

Diffuse reflection.

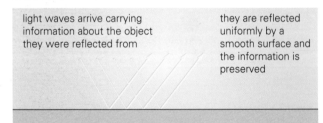

Regular reflection.

Mirrors

As you can see from the diagram of regular reflection, light reflects from a mirror in a very predictable way. If the mirror is perfectly smooth then light will always be reflected from the mirror at the same angle that it arrived. These angles are conventionally measured against a line drawn at right angles to the mirror. This line is called the **normal**. The angle a ray of light arriving at the mirror makes with the normal is called the **angle of incidence**. The angle that the reflected ray makes with the normal is called the **angle of reflection**. These two angles are always equal:

angle of incidence (i) = angle of reflection (r)

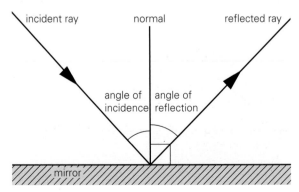

A reflected ray makes the same angle with the normal as the incident ray.

Images in plane mirrors

When you look in a mirror you are seeing light which has been reflected from the mirror. This means that the image produced will have certain characteristics. The most obvious characteristic is that when you look into a plane (flat) mirror your image appears to be *as far behind the mirror as you are in front of it*. This happens because the light rays entering your eye are following a path that they would have taken, had they travelled in a straight line from behind the mirror. Your brain interprets this by 'seeing' the image behind the mirror.

The diagram shows another characteristic of an image in a plane mirror – it is **laterally inverted**. This means that the right-hand side of the object being reflected appears on the left-hand side of the image. Try writing your name on a piece of paper and holding it in front of a mirror – you will see what effect lateral inversion has!

The image in a mirror is an illusion caused by the way a mirror makes light behave. There is no image behind the mirror in reality. Because the image is not real it is called a **virtual image**.

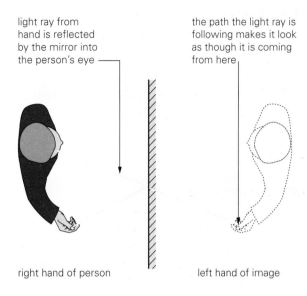

The image appears to be behind the mirror and is laterally inverted.

Drivers rely on the image in a mirror to judge the distance from them of cars behind.

QUESTIONS

1a What is a *normal* line?
b What is the rule of reflection?

2a Draw a diagram of a ray of light arriving at a mirror with an angle of incidence of 30°.
b Now add the reflected ray to your diagram.

3 Explain why when you look in a mirror you can see yourself but when you look at a piece of paper you cannot.

4a List the characteristics of the image formed in a plane mirror.
b How does *your* image differ from how your friends see you?

6.7 Refraction of light

In 6.2 you saw how water waves are refracted when they change speed. As light is a wave then it will also be refracted when it changes speed. Although light always travels at 3×10^8 m/s in a vacuum or in air, when it travels in a medium such as glass, Perspex or water its speed is lower. Typically the speed of light in such a material will be more like 2×10^8 m/s, but the exact value depends on the medium *and* the wavelength of the light.

This change in speed will cause a ray of light to be refracted as it passes between different media. You can see this happening if you shine a ray of light from a ray-box into a glass block at non-normal incidence. As the ray of light enters the glass it 'bends' towards the normal – the angle the refracted ray makes with the normal (the **angle of refraction**) is smaller than the angle the incident ray makes with the normal (the **angle of incidence**). If you increase the angle of incidence then the angle of refraction will also increase, but it will always be smaller than the angle of incidence. As the ray leaves the glass block it bends away from the normal again, emerging parallel to the incident ray, but displaced.

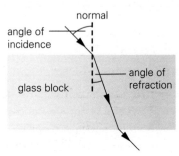

As the ray of light enters and leaves the glass block, it is refracted.

In general:

- As light enters a *more dense medium*, such as glass, from a less dense medium, such as air, it will *always bend towards the normal*.

- As light enters a *less dense medium*, it will always bend *away from the normal*.

This change of direction can be explained by thinking of the light ray entering the glass block in just the same way as water ripples entering a shallow region in a ripple tank (see 6.2). As the wavefront of the light reaches the glass block it slows down. When the light ray hits the block at an angle, the first part of the wavefront to arrive at the glass block will slow down while the rest continues at the same speed. This causes the direction of the ray of light to change. The change of direction on leaving the glass block can be similarly explained.

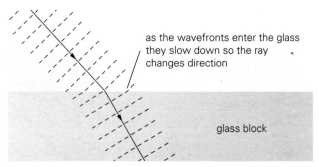

The ray of light behaves like water waves in a ripple tank – as it slows it changes direction.

Prisms

A prism is a block of glass, usually with a triangular cross-section. When light passes through a triangular prism the refraction on entering and leaving causes the emergent ray to be **deviated** – in a different direction– from the incident ray.

If you pass a ray of white light from a ray-box through a triangular prism you may see another effect. The prism splits the light into the seven colours that, mixed together, make up white light. These are shown at the top of the next page.

The effect of refraction is increased by a triangular prism.

This is called **dispersion** and is due to the different colours of light being refracted (and hence deviated) by different amounts.

Each different colour of light has a different wavelength. Red light has a wavelength of about 7×10^{-7} m, violet light has a wavelength of about 4×10^{-7} m, and the other colours have wavelengths between these two extremes. Red is deviated least by a prism and violet most.

The amount of refraction by a medium clearly depends upon the wavelength of the wave – longer wavelengths are refracted less than shorter wavelengths. This means that longer wavelengths are slowed down less when they enter a more dense medium.

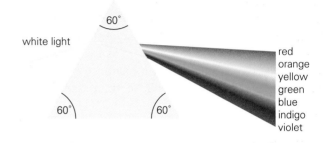

A prism will separate out the spectrum of colours that make up white light.

The spray at the bottom of a waterfall can disperse light.

QUESTIONS

1. Write out the colours of the spectrum of white light in order of increasing *frequency*.

2. When the light passes from one medium into another, what is the only angle of incidence that will give an *equal* angle of refraction?

3. Copy out the following sentences and fill in the blanks:

   As light travels from a _____ dense medium to a _____ dense medium it will bend away from the _____ . As it travels from a less dense medium to a more dense medium it will bend _____ the normal.

4. Explain why light refracts as it moves from air to glass.

5. Why does a triangular prism separate the spectral colours but a rectangular glass block does not?

6.8 Total internal reflection

If you have ever been underwater in the swimming pool and wondered why the surface of the water appears to be reflective, like a mirror, then you have seen **total internal reflection** in action.

When a ray of light has a path which takes it from a dense medium to a less dense one – for example, from water to air – then it will be refracted away from the normal. The angle of refraction will in this case be *greater* than the angle of incidence. It follows that there will be an angle of incidence which will produce an angle of refraction of 90°. The refracted ray will pass along the boundary between the air and water, grazing the surface. This angle of incidence is called the **critical angle**.

The surface of water seen from below looks silvery.

If the angle of incidence is made greater than the critical angle then the angle of refraction would have to be greater than 90°. This means that the ray of light would be refracted back into the water! What actually happens is that beyond the critical angle the light is *reflected* back into the water. This behaviour is called total internal reflection, because all of the light is reflected. (*Some* light is always reflected from a boundary between a dense and a less dense medium.)

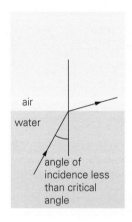 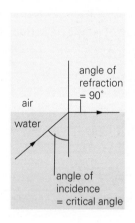 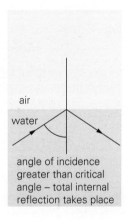

As the angle of incidence increases, so does the angle of refraction. Beyond the critical angle the light cannot escape from the water and is totally reflected.

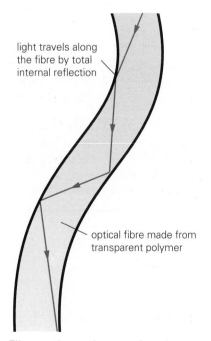

Fibre optics make use of total internal reflection.

Fibre optics

Modern communications depend upon total internal reflection of light. The 'Information Super-Highway' is a computer-based network which, it is planned, will allow very fast transfer of huge amounts of data between homes, offices and network servers. Some technologists believe that this will bring about a new industrial revolution, causing a major change to the way we live and work. This 'Super-Highway' requires a very good quality network of connecting cables, capable of carrying millions of pieces of information every second. The traditional way of carrying information of this sort is to use electrical signals in copper wire, but this method is not capable of transmitting the quantity of information needed. The solution is to use **optical communication**. Information can be **digitally coded** into a series of pulses of light (rather like Morse code). These pulses of light are emitted by tiny lasers (whose 'light' is actually in the infra-red region of the electromagnetic spectrum) and are then sent along a fibre optic cable. The fibre traps all the light inside using total internal reflection.

The pulses of light are decoded at the other end of the cable. All kinds of information – speech, music, pictures – can be sent at the speed of light! Telephone links, cable TVC transmissions and computer networks already use fibre optic cable.

There are several advantages of optical communication over electrical transmission in copper cables:

- A fibre optic cable about 2 cm in diameter can carry, for example, as many as 500 000 telephone conversations simultaneously! By comparison, a much thicker copper cable could carry perhaps 1000.
- Fibre optic cables are much lighter and so easier to lay than copper cables. They are also cheaper.
- Optical signals do not suffer from electrical interference, there is no distortion and less reduction of the signal over long distances.

Fibre optic cable.

Optical fibres are also used in medicine. An 'endoscope' is a long flexible instrument consisting of a bunch of fibres which can be inserted, for example, into the digestive tract. When light is passed down the instrument it is reflected back up the fibres and an image can be obtained of the area inside the body.

The modern technique of 'keyhole' surgery uses a much thinner version of the endoscope. One use for this device is in heart surgery where the probe is passed through a patient's vein right into their heart to inspect the valves for any damage or obstructions. This has enormous benefits to the patient compared with traditional surgery because it can be carried out under a local anaesthetic and requires only a very tiny entry wound.

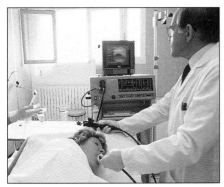
An endoscope uses optical fibres to see inside a patient's body.

Optical instruments

Another application of total internal reflection is in optical instruments such as binoculars and periscopes. In these devices it is necessary to change the path of light using reflectors.

Why not use mirrors? When light is reflected in a mirror, most of it is reflected by the silvered bottom surface of the glass, but some is also reflected by the top surface of the glass. This leads to multiple images very close together, and can cause the image to be blurred.

The solution to the problem is to use prisms. The type of prism used in binoculars has a right angle and two 45° angles. The prisms are arranged so that light enters at 90° to one rectangular surface. It is then totally reflected at each of the other two surfaces. There is no possibility of multiple images.

The production of a magnified image in binoculars requires a long light path. Reflectors are needed to keep the instrument short enough to handle.

QUESTIONS

1. What is meant by the term *critical angle*?
2. Explain total internal reflection.
3. How does an optical fibre make use of total internal reflection?
4. Why are prisms rather than mirrors used in periscopes?

6.9 Diffraction

When you make plane waves in a ripple tank travel through a gap or around an obstacle, they 'spread out'. This spreading out is called **diffraction**. If the gap or obstacle is large compared with the wavelength of the wave, then the waves only spread out by a small amount. As it approaches the size of the wavelength, however, the wave spreads out more and more until a circular wavefront emerges.

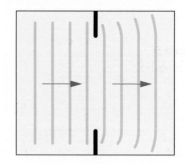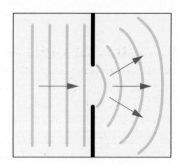

Diffraction increases as the gap approaches the size of the wavelength.

The photograph shows what happens to sea waves as they pass through a harbour entrance. You can see the waves diffracting as they enter the harbour.

Diffraction is not limited to water waves – it is an effect shown by all types of wave with an extended wavefront, whether transverse or longitudinal. The fact that electromagnetic radiation can be diffracted provides confirmation that it travels as a wave.

Waves diffract as they enter a harbour.

Diffraction of light

In order to see diffraction of light you must pass it through a gap that is only thousandths of a millimetre across, because of the requirement that the gap must approach the wavelength of the wave. This means that you rarely see light being diffracted. If you shine a laser beam through a very narrow slit, however, you can see it spread out and form into light and dark bands.

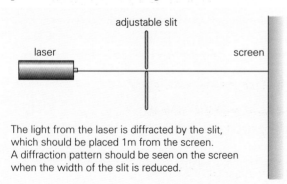

The light from the laser is diffracted by the slit, which should be placed 1m from the screen. A diffraction pattern should be seen on the screen when the width of the slit is reduced.

A laser beam shining directly onto a screen.

A laser beam after diffraction through a narrow slit.

Diffraction of radio waves

The very large wavelength of some radio waves means that they are very easily diffracted by ordinary obstacles. This explains why it is possible to receive long-wavelength transmissions even if you live in the shadow of hills which block out line-of-sight transmissions.

You can demonstrate the diffraction of radio waves yourself by using 3 cm microwaves and an adjustable gap formed by two metal plates.

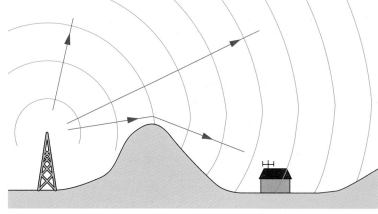

Some radio waves have large enough wavelengths to be diffracted by hills and mountains.

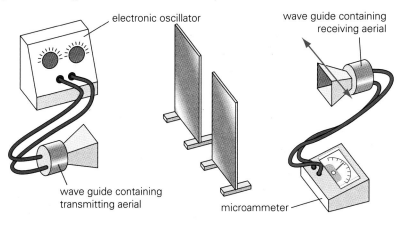

Diffraction of the microwaves can be shown by moving the receiver from side to side.

Diffraction of sound waves

Most sounds we listen to are in the frequency range of a few hundred hertz to a few thousand hertz. If 1000 Hz is taken as an average frequency of sound then its corresponding wavelength in air is about 30 cm. This means that sound waves will be diffracted by gaps and obstacles of roughly that size. The most common gap of this sort is a doorway, and indeed sound waves are diffracted by doorways. You can try a simple experiment. Ask somebody to talk to you through a doorway. As they do so, stand out of sight on one side of the doorway about 1 metre from the wall. Walk across the doorway. What happens to the loudness of the sound?

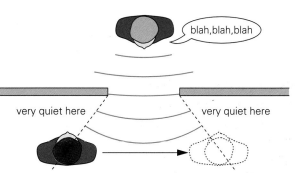

Sound waves are diffracted as they pass through a doorway because the wavelength is of the same order of magnitude as the width of the door.

QUESTIONS

1a What is meant by *diffraction*?
 b What is the condition for maximum diffraction of a wave as it passes through a gap?

2 X-rays are diffracted as they pass through the layers in between atoms in a crystal. What does this tell you about the wavelength of X-rays compared to the interatomic spacing in a crystal?

3 Why are people who live in mountainous terrain often unable to pick up VHF (very high frequency) transmissions?

CHAPTER 7: SOUND

7.1 How are sounds produced?

Sound is enormously important to us. Try to think for a moment about all the times information is given in the form of sound. For example, learning at schools can be very difficult for deaf people, and they cannot communicate by ordinary telephone either. Apart from these practical considerations, just think how much enjoyment people get from listening to music.

Sounds are produced whenever an object vibrates. If you play the guitar you can see that when you pluck a string it vibrates up and down. If you strike a tuning fork you will see the prongs vibrate.

As the prongs move outwards they push the air particles together, compressing the air and so increasing the air pressure slightly. As the prongs move back again the air particles are able to become more spread out, and the air becomes 'rarefied', i.e. the air pressure drops slightly. These successive **compressions** and **rarefactions** move outwards from the tuning fork as a longitudinal wave, like the longitudinal wave on the slinky spring that you saw in 6.1. As with all waves, energy is transferred. There is no overall displacement of the air – as the wave passes, individual particles of air move backwards and forwards.

The vibrations of a guitar string or a tuning fork are regular, and will produce a regularly repeating sound wave. Other vibrations, such as the 'clash, clash' of cymbals, are far from regular. The sound still travels away as a series of compressions and rarefactions, but these do not form a regular wave. This type of sound may be referred to as 'noise'.

We get a lot of enjoyment from different sounds.

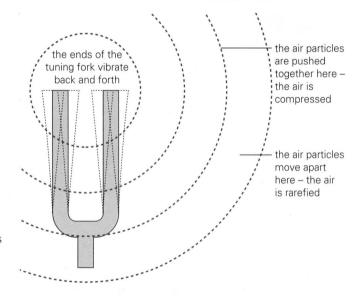

As the tuning fork vibrates it produces longitudinal sound waves in the air.

Sound needs a medium

Sound travels, then, by making the particles in the air oscillate. When the oscillations reach your ear, your eardrum is made to vibrate. The vibrations are passed through to the inner ear to be picked up by nerve endings which transmit corresponding impulses to your brain.

The medium does not have to be air. Whales and dolphins can communicate over many kilometres under water by 'singing' to each other. This is possible because sound waves travel very well in water.

The sound of singing whales carries over huge distances.

If there are no particles, sound waves cannot be made – sound cannot travel in a vacuum. You can prove this with a simple experiment. If you set an electric bell ringing inside a sealed bell jar you will be able to hear the sound of the bell. If you then pump all the air out of the bell jar using a vacuum pump, the bell becomes quieter and quieter until, in the end, when there is no air left you cannot hear the bell at all. If you open the valve, air rushes in and you hear the sound again.

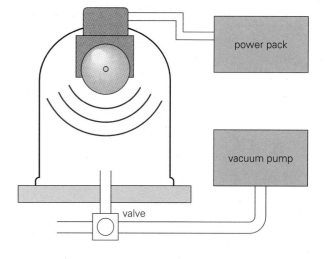

This experiment shows that sound needs a medium to travel through.

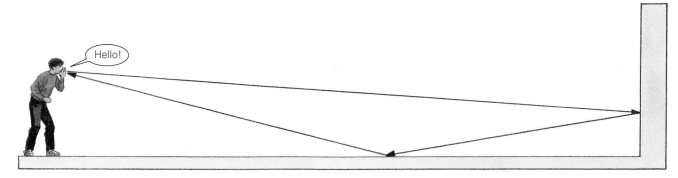

Echoes are caused by sound reflecting from a hard flat surface.

Echoes

Echoes are heard when sound waves are reflected from a hard flat surface. You will probably have experienced echoes when you have been in a sports hall or in a cave. These places make very clear echoes because the reflecting surfaces are very hard and are a long way from the source of the sound, so making the time delay between the original sound and its echo quite long.

Where the distance to the reflecting surface is small then the time delay for the echo is very small, and will probably overlap the original sound. This is called **reverberation**. There is a lot of reverberation in small rooms with hard surfaces, such as a bathroom. If you sing in the bath, the hard surfaces reflect nearly all the sound energy and this makes your voice sound much better! You hear a very different 'quality' of sound if you sing in the living room. Here the carpets, curtains and soft furniture absorb most of the energy of the sound and reflect only a tiny part of it.

QUESTIONS

1 How does a vibrating guitar string make a sound?

2 The Moon has no atmosphere. Would there be any point in playing a guitar on the Moon? Explain your answer.

3 What does the word *reverberation* mean?

4a People who design concert halls spend a lot of money getting the degree of reverberation correct. Why do you think this is?

b What could a designer do to reduce reverberation if there is too much?

7.2 The speed of sound

Sound waves take time to travel from a source of sound to your ear. The speed of the sound wave depends on the properties of the medium through which it travels. If the particles in the medium are spaced widely apart, for example, the oscillations will not be passed on as quickly as if the particles are packed closely together. In a dense solid, such as steel, the molecules are packed closely and have strong permanent linkages (or **bonds**) between them. This means that the speed of sound will be very high – it is about 6000 m/s in steel. By contrast, the molecules in gases are far apart and have no linkages. The speed of sound in air is only about 330 m/s.

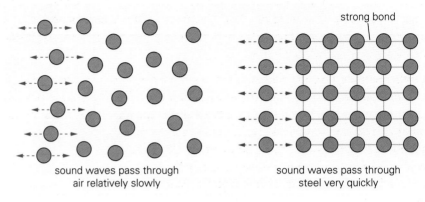

The speed of sound depends on the medium.

Refraction of sound

Just like light and water waves, sound waves can be refracted. Remember that refraction occurs when a wave changes direction as a result of a change in speed. This change in speed can happen when a sound wave moves from one medium to another. For example, sound travels more slowly in carbon dioxide than it does in air. This means that a balloon filled with carbon dioxide can be used to focus sound waves.

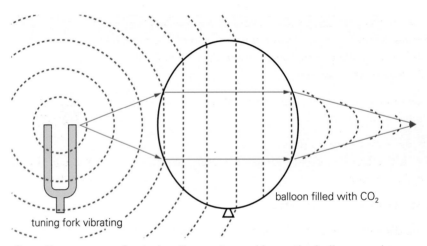

Sound waves are refracted as they enter and leave the balloon, and are focused to a point.

Measuring the speed of sound in air

Even though, at 330 m/s, the speed of sound in air is relatively slow compared to its speed in steel, it is still fast enough to cause some difficulties if you wish to measure it. The usual way to measure the speed of a moving object is to time it over a known distance. How can this be accomplished in the case of moving sound waves? In order to make the travel measurable, the distance must be large. A distance of 100 m would produce a time of travel of about 0.3 seconds. A sound loud enough to be heard 100 m away, for example a starting-pistol shot, must be made on a given visual signal. On the same signal the person standing at 100 m distance should start a stop-watch and then stop it again when the gunshot is heard.

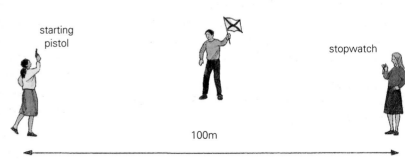

One way of measuring the speed of sound is to time it over a known distance.

The obvious problem with this method is the speed of human reactions. Most people will take in the region of 0.1 seconds to react to an event. Both the person firing the shot and the timer have to react to the initial signal, then the timer has to react to hearing the shot. The result of this experiment will probably tell you more about the reaction times of the people carrying it out than it will about the speed of sound!

Another method is to use echoes. If you stand 20 m, say, from a large wall and bang two pieces of wood together, you will hear the echo of the noise you made a split-second later. The time delay of the sound is the time it has taken for the sound to travel to the wall and back; in this case it would take about 0.1 seconds for the echo to return. This time is much too short to measure, but you can increase the measured time by clapping in time with the echoes to set up a steady rhythm: *clap, echo, clap, echo.*

If it takes 0.1 seconds for the echo to return then to make a steady rhythm you must wait a further 0.1 seconds before clapping again. The time between two claps will therefore be 0.2 seconds. This is the time needed for the sound to make two return journeys.

You could time, say, 30 claps. As an example, this time might be 7.3 seconds.

time taken to clap 30 times
  = time taken for sound to make 60 return journeys
  = 7.3 seconds

distance to wall = 20 metres

total distance travelled by sound in 7.3 seconds
  = length of one return journey × number of journeys
  = (2 × 20) × 60 = 2400 m

$$\text{speed of sound} = \frac{\text{distance}}{\text{time}}$$
$$= \frac{2400}{7.3} \approx 330 \, \text{m/s}$$

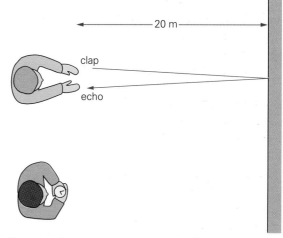

You can also measure the speed of sound using echoes.

QUESTIONS

1 A group of students decide to measure the speed of sound by firing a starting pistol and timing the sound over 100 m. They take 5 readings of the time taken so that they can take an average:
0.2 s, 0.24 s, 0.39 s, 0.15 s, 0.3 s

Calculate the speed of sound from this data. How reliable do you think the result is? Explain your answer.

2 Another group of students decide to use the clap/echo method. One of the group stands 60 m from a wall and claps in time with the echo. Another student times the experiment and finds that it takes 7.0 seconds for the student to clap 21 times. The experiment is repeated twice more with the following results: 6.8 s, 7.1 s. Calculate the speed of sound from this data.

3 The speed of sound in water is 1400 m/s. Explain why sound travels faster in water than in air.

7.3 Picturing sound waves

From what you have learnt about sound so far you know that sound travels through the air as a variation in air pressure, or a **pressure wave**. You saw in chapter 6 that waves can be described by their frequency, wavelength and amplitude, but it is not very easy to see how these characteristics relate to a pressure wave.

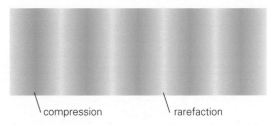

Sound waves travel through the air as a series of compressions and rarefactions.

Using a cathode ray oscilloscope

To get a useful picture of a sound wave you can use a microphone and a cathode ray oscilloscope. The microphone is used to convert the sound wave into electrical signals (see 3.4). The oscilloscope display shows how that signal changes.

The microphone is connected to the 'Y-input' of the oscilloscope (see 1.8). The changing voltage from the microphone then makes the oscilloscope trace go up and down as it moves across the screen. If a 'pure' note (i.e. a regular sound wave of a single frequency) from a tuning fork is played into the microphone, a regular transverse waveform is seen on the screen.

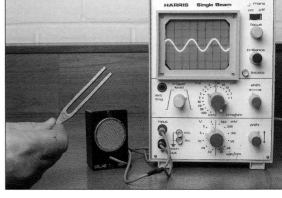

A cathode ray oscilloscope enables you to 'see' a sound wave.

Alternatively, a signal generator can be used instead of a tuning fork and microphone to generate a pure note. The effect on the trace of varying the frequency and the amplitude can then be seen.

The illustration shows the pressure wave of a pure note compared with its oscilloscope display.

The high points on the oscilloscope trace correspond to the compressions and the low points to the rarefactions.

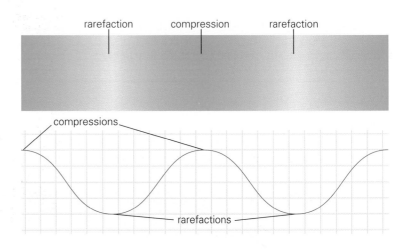

Pitch and frequency

If you twang a ruler on the edge of a desk you will see that if the protruding length of the rule is shorter it vibrates more quickly. Notice also that as the frequency of the vibrations increases, the **pitch** of the sound you hear gets higher. You can compare the traces of a low-pitched pure note and a high-pitched one on an oscilloscope. You will see that the higher-pitched note has a shorter wavelength.

The normal range of sound frequencies that a young person can hear, called the **audible range**, is from 20 Hz to 20 000 Hz. As you get older the upper limit of this range decreases.

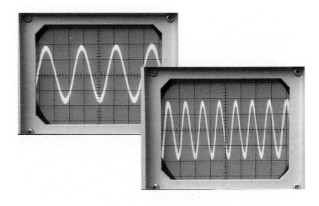

A low-pitched note (top trace) has a lower frequency than a high-pitched note.

Loudness and amplitude

The amplitude of a wave is a measure of the energy it carries. A sound wave carrying more energy has bigger differences in pressure between the compressions and rarefactions than a quiet sound – its amplitude is greater (see 6.1). The greater pressure variations at the ear are interpreted by the brain as a louder sound.

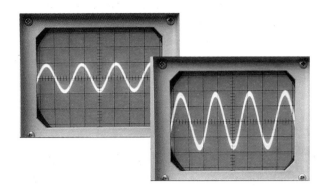

A quiet sound (top trace) has smaller amplitude than a loud sound.

Louder and louder

The conversion of sound signals into electrical signals in a microphone enables them to be **amplified** by an electronic circuit and fed into a communications network or passed directly to a loudspeaker. A loudspeaker converts the electrical signals back into a sound wave (see 3.3).

Prolonged exposure to very loud sounds can damage the ears, particularly their sensitivity to sounds in the upper part of the audible range. Ear defenders are advisable for people operating very noisy machinery.

Persistent or repetitive unwanted noise at any level can cause other problems such as stress and difficulty in concentrating. This is referred to as 'noise pollution'. It can be reduced by various means, including:

- double glazing, which acts as a barrier to sound waves
- thick curtains and carpets, which absorb much of the energy of sound waves
- acoustic tiles, which 'trap' the sound waves in air pockets.

If you want to make yourself heard, you have to make yourself loud!

QUESTIONS

1 How does the pitch of a note relate to its frequency?

2a What does *audible range* mean? What is the normal human audible range?
b How could use of a personal stereo affect a person's audible range?

3 Draw an oscilloscope trace of two pure notes:
a of the same loudness, but the second one with a lower pitch than the first
b of the same pitch, but the second one quieter than the first.

4a Explain, with the help of a diagram, what the amplitude of a pressure wave is.
b Why does the loudness of a sound depend on the amplitude of the wave?

7.4 Ultrasound

Sounds with frequencies above the normal human upper audibility limit of 20 000 Hz are called **ultrasound** or **ultrasonic waves**.

Different animals have different audible ranges. Dogs can hear much higher frequencies than we can. Some people use so-called 'silent' dog whistles to call their dogs. These whistles are not silent but whistle at an ultrasonic frequency that is within the audible range of dogs.

Ultrasound echo-location

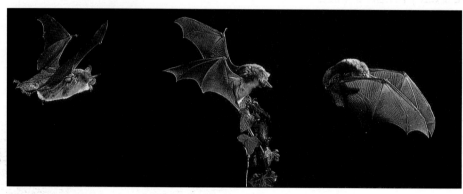

Ultrasound enables bats to fly with deadly accuracy.

Bats make use of ultrasound in a very special way. They fly at night, and so their sight is almost useless to them. A bat's eyesight is of such little importance to it that most species of bat are gradually losing the use of their eyes altogether. To enable them to navigate at night they have developed an ability to 'see' using echo-location. They emit a series of ultrasound squeaks which bounce back from obstructions. The very short wavelength of the waves means that they spread out very little. The directional nature of the signal, along with the time taken for the echoes to return, allows the bat to build up a very accurate picture of the area it is flying through and of the location of its prey.

Depth-testing using ultrasound

Using a copy of the bats' technique, many boats are now equipped with devices that use ultrasound echo-location (or 'sonar') to find out the depth of the water they are in. The echo-sounder sends out ultrasound pulses and a monitor times how long they take to return. Again, the highly directional nature of ultrasound allows a very accurate determination of depth to be taken.

Ships use ultrasound echo-location.

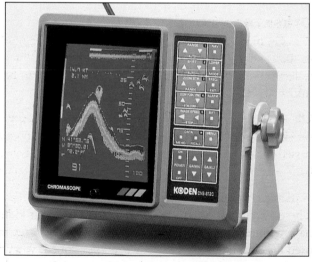

The readout on the ship's bridge gives the crew an accurate map of the sea bed.

Scanning unborn babies

It is very useful for doctors to be able to see an unborn fetus in its mother's uterus. The number of fetuses present can be seen and the fetal position and growth rate checked, and some abnormalities can be discovered. X-rays cannot be used for this purpose because they might damage the fetus. Ultrasound scanners are safe and have the added advantage that they are able to distinguish between different types of tissue in a way that X-rays can't.

The hand-held ultrasound 'transducer' sends out a very high-frequency ultrasound signal produced by an electronic circuit. As the waves pass through different tissue types, different amounts of the energy are reflected back to the transducer.

The reflected signals and the time they take to return are then processed by a computer and enable an image of the fetus in the uterus to be produced.

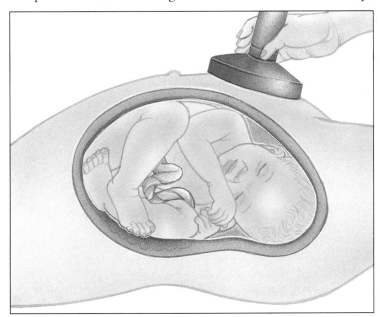

Ultrasound scanning works by analysing the reflections of signals sent out by a transducer.

It isn't only pregnant women who are scanned. This sow is expecting piglets!

Other uses for ultrasound

Flaws in metal joints and castings can be found by a method similar in principle to the scanning of fetuses. Ultrasound passes through metal. A flaw will cause reflections of an incident ultrasound wave, which can be analysed to give information on the location and nature of the defect.

Ultrasound can also be used for cleaning delicate mechanisms without having to take them apart. High-energy (i.e. high-frequency) ultrasonic waves are passed through a bath of liquid. The mechanism to be cleaned is put in the bath and the vibratory energy from the waves removes the dirt.

A particularly painful condition that some people suffer from is kidney stones. These are crystals of urea that can form inside your kidneys. Traditionally they have required surgery to remove them. Recently a technique similar to that used for cleaning machinery has been developed, whereby the kidney stones are shattered by highly-focused high-energy ultrasound waves. Once shattered they can be passed, relatively painlessly, out with the urine.

QUESTIONS

1. List five different uses for ultrasound.

2. Why is ultrasound used for pre-natal scanning rather than X-rays?

3. Why is it important when ultrasound is used for echo-location that it does not spread out much? Draw a diagram to help your explanation.

7.5 Seismic waves

Why do earthquakes occur?

Earthquakes can cause massive devastation. They are a result of the structure of the Earth's surface layers. The continents sit on top of enormous rock 'plates', known as **tectonic plates**, which make up the crust of the Earth. These plates are able to move very slowly, at about 1 cm per year.

A major earthquake devastated large areas of Kobe, Japan.

The plates are moving apart in the middle of the Atlantic Ocean. This allows molten rock to be forced up from beneath the crust, making the sea bed of the Atlantic bigger all the time. A series of islands have been created through this process, including Iceland and The Azores.

At the edges of the Pacific Ocean the plates are being forced together, so one plate is forced below the other. This process puts an enormous amount of stress on the rock that makes up the tectonic plates. Quite frequently, the energy stored in the rocks will release itself suddenly as the plates jerk past each other. This release of energy is an earthquake. The site of the earthquake is called the **focus** and the nearest point to the focus on the Earth's surface is called the **epicentre**.

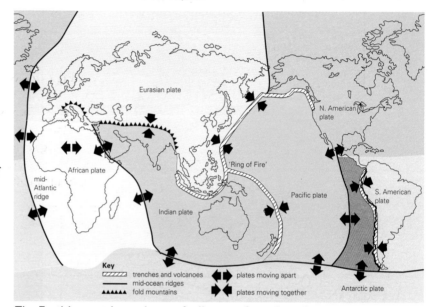

The Earth's crust is made up of a jigsaw of moving plates.

Most of the destruction of buildings and other structures occurs as a result of huge ground movements close to the epicentre. Earth tremors, however, can be felt far from the epicentre. This is the result of waves of energy travelling away from the focus of the earthquake through the Earth. The waves are called **seismic waves**. There are two main types of seismic wave:

- primary or **P waves** are *longitudinal* and can travel through solids and liquids
- secondary or **S waves** are *transverse* and can only travel through solids.

Both P and S waves spread out very rapidly from the focus of the earthquake in all directions through the Earth. They can be detected and measured using a **seismometer**.

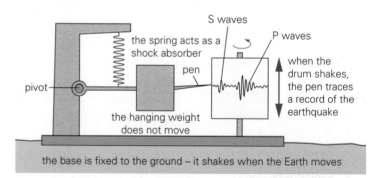

A seismometer is used to make a recording of earthquake waves. A seismogram shows that P waves arrive first and cause the first tremors. Later S waves arrive and cause a second set of tremors.

It can be seen from a seismometer trace (or 'seismogram') that the P waves arrive before the S waves. This shows that P waves travel faster than S waves.

The different speeds of S and P waves through the crust allow the distance of the focus of an earthquake from the recording station to be calculated precisely. If these calculations are made from a number of different stations then it is a simple matter to work out where the focus and hence the epicentre of the earthquake is.

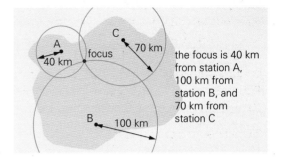

the focus is 40 km from station A, 100 km from station B, and 70 km from station C

Finding the epicentre of an earthquake.

The structure of the Earth

Seismic waves travel through the Earth in curved paths – they are refracted in material of varying density, because of a change in speed. Both P and S waves travel faster in denser rock. At a boundary between different types of material, their paths change abruptly due to a change of density and some energy is also reflected.

Studying seismograms recorded at different points on the Earth's surface has enabled scientists to determine the paths the P and S waves have taken from the focus and hence to build up a picture of the structure of the Earth.

Putting together the evidence from seismic waves, it appears that the Earth is covered in a thin solid layer – the **crust**. Once seismic waves have travelled 25–40 km through the crust they suddenly speed up and change direction, suggesting that the density of the rock has increased. This marks the boundary between the crust and the next layer, called the **mantle**.

Evidence suggests that the density of the mantle increases with depth and that this layer extends almost halfway to the centre of the Earth. At about 3000 km below the surface, P waves slow down and S waves stop altogether, creating 'shadow-zones' on the surface where no S waves are received. Since S waves cannot travel through liquids this gives evidence that the **core** of the Earth is at least partly liquid. In fact the outer part is liquid and the central part is solid.

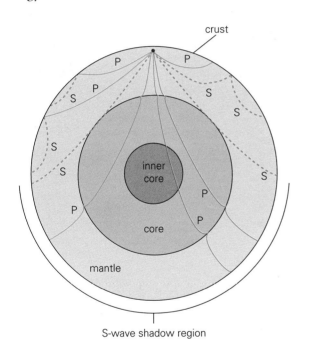

Study of seismic waves shows the Earth is made up of different layers.

QUESTIONS

1 What are the differences between P waves and S waves?

2 Why do seismic waves bend outwards as they travel through the mantle?

3 Explain why one seismometer reading cannot locate the epicentre of an earthquake, but readings from three widely spaced seismometers can.

4 What is the evidence for the liquid nature of the Earth's outer core?

5 A seismic recording station records the arrival of S waves 5.0 seconds after the arrival of P waves. If the P waves travel at 20 km/s in the underlying rock and the S waves at 5 km/s, how far from the focus of the earthquake is the recording station?

SECTION C: QUESTIONS

Take the speed of electromagnetic waves in a vacuum as 3×10^8 m/s and the speed of sound in air as 330 m/s.

1. Explain the meaning of the words in *italics* in the following paragraph.

   There are two types of *wave*, *transverse* and *longitudinal*. In some respects they behave differently from each other but they can both be described in terms of their *wavelength*, *frequency*, *speed* and *amplitude*.

2. a. What is the equation that links the speed of a wave to its frequency and its wavelength?
   b. A ripple tank bar oscillates with a frequency of 20 Hz. The ripples produced have a wavelength of 3 cm. What is the speed of the ripples in the ripple tank?
   c. Radio 1 FM is broadcast in a frequency band of 97.6–99.8 MHz. Work out the wavelength range of the waves.
   d. Red light has a wavelength of 7×10^{-7} m when it is travelling in a vacuum. Work out the frequency of the electromagnetic oscillation.

3. Make an accurate copy of these wavefronts in a ripple tank and complete the diagram to show where the ripples go.

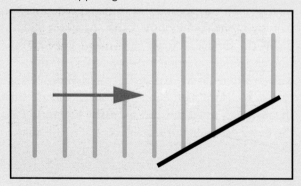

4. Which of the objects at points A, B, C, D and E would you see in the mirror if you were at X?

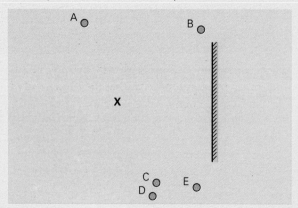

5. When a water wave changes speed it may also change direction.
   a. What is this behaviour called?
   b. What causes a water wave to change speed?
   c. Copy out and complete these two diagrams of wavefronts in a ripple tank.

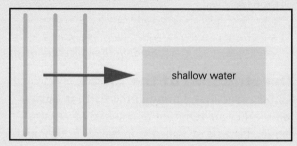

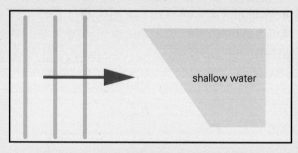

6. The diagram shows some glass blocks with light rays passing through them. In each case choose the correct path A, B, C or D.

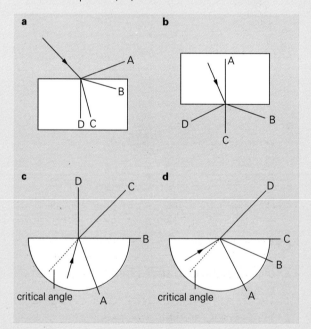

7. a. Explain, with the help of a diagram, how an optical fibre makes use of total internal reflection.
   b. Describe how optical fibres are used in communications systems, and give the advantages of such a system over electrical transmission.

8   When you shine white light from a halogen bulb through a glass prism, it separates into different colours.
   a   What does this dispersion tell you about the speed of different wavelengths of light in glass?
   b   Where would you expect any infra-red radiation from the bulb to be in relation to the visible spectrum produced, and how could you detect its presence?

9   Which parts of the electromagnetic spectrum would you use for the following purposes?
   a   making a piece of toast
   b   getting a good suntan
   c   finding out if you had a broken leg
   d   transmitting information via a geostationary satellite
   e   treating cancer

10 The diagram shows four different traces of sound waves on an oscilloscope.

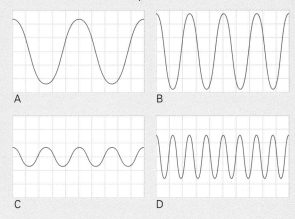

   Which of A, B, C or D shows the:
   a   highest-pitch sound?
   b   loudest sound?
   c   lowest-pitch sound?
   d   quietest sound?

11a An ultrasound transducer 2 cm across will produce a very directional 'beam' of waves. If the same size device were used to produce audible sound waves then the sound would spread out. Explain this using the idea of *diffraction* of waves.
   b   Why is ultrasound, and not sound of audible frequencies, used for echo-location?
   c   A fishing boat's sonar equipment sends out an ultrasound signal to locate a shoal of fish. The reflected signal is received 0.9 s later. How far away is the shoal? (The speed of sound in water is 1400 m/s.)

12 When large buildings are demolished using explosives, the event usually attracts a large crowd. For safety reasons the crowd has to stand a long way away from the building that is being blown up. It appears to the people watching that the building starts to collapse before the explosive is detonated. How do you explain this? The building starts to collapse 1.5 seconds before the crowd hear the explosion. How far away are the people standing?

13 Explain the meaning of the words in *italics* in the following paragraph.

   Earthquakes are produced when large amounts of energy stored in the *Earth's crust* as a result of *tectonic plate* movement are suddenly released. The energy travels from the *focus* of the earthquake to the surface in the form of two types of seismic waves, *P waves* and *S waves*. The different speeds of these waves allow the *epicentre* of the earthquake to be located.

14a Draw a labelled diagram of the structure of the Earth.
   b   What do the paths of these seismic waves tell you about the density variation within layers 2 and 3, and the density changes at the boundaries between layers 1 and 2, and layers 2 and 3?

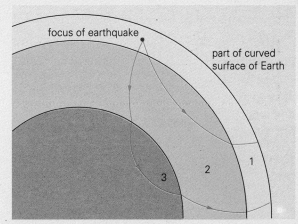

   c   What evidence from the detection of seismic waves supports the view that the outer core of the Earth is liquid?

CHAPTER 8: MOVING ENERGY AROUND

8.1 Where does energy come from?

Energy is needed to make things happen. It exists all around us in many different forms, some of which are more convenient for us to use than others. Energy is continually being converted from one form to another. Some conversions occur naturally, for example, when the wind blows; and others only when we allow them to, as for example when we switch on a torch.

Energy is measured in joules (J), the unit being named after the nineteenth-century British physicist James Joule, who investigated energy conversions.

James Joule, 1818–1889

Different forms of energy

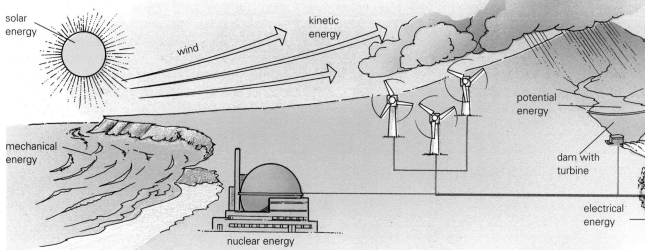

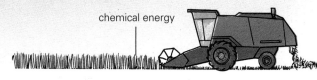

Kinetic energy is the energy an object has as a result of its movement. The faster it moves, the more kinetic energy it has.

Potential energy is stored energy that an object has because of its position. The higher an object is, the more **gravitational potential energy** it has. Energy stored in a stretched or compressed spring is **elastic potential energy**.

Sometimes it is convenient to use the term **mechanical energy**, which is the amount of kinetic *and* potential energy that something has. For example, we might talk of the mechanical energy of a moving piston.

Chemical energy is released when a chemical reaction, such as combustion, takes place. Substances like coal which are able to release a large quantity of energy when they burn are used as **fuels**. Electric cells are also a store of chemical energy.

Internal energy or **thermal energy** is the energy that something has as a result of the movement of its molecules. When something is heated, its internal energy increases and its molecules move faster.

Electrical energy may be the energy of moving electric charges in a conductor, or the potential energy stored due to an excess of static electric charge.

Sound energy is the energy of vibrating molecules due to the passing of a sound wave.

Nuclear energy is released from the nucleus of an atom when a nuclear reaction takes place.

Solar energy consists of **light energy** and **radiant (infra-red) energy** from the Sun. These are forms of electromagnetic radiation.

Most of our energy comes from the Sun

The Sun is our main provider of energy, supplying over 5000 times more energy to the Earth's surface each second than all other sources (such as radioactive rocks) combined. It warms the ground, the oceans and the atmosphere, providing the energy for ocean currents, wind and other weather conditions (see 8.4). Much is reflected or re-radiated back into space.

About 0.02% of the solar energy that reaches the Earth's surface is stored as chemical energy by plants, as a result of photosynthesis (see 9.5). Plants are eaten by animals, passing the energy on along the food chain. This stored chemical energy in our food is released by our bodies to give us the energy needed for all body functions and all activity.

Some plant material, notably wood, is used as a fuel. **Fossil fuels** – coal, oil, natural gas and peat – are major energy sources. They were formed millions of years ago from the remains of plants and animals. The chemical energy stored in the fuel is converted to thermal energy on burning.

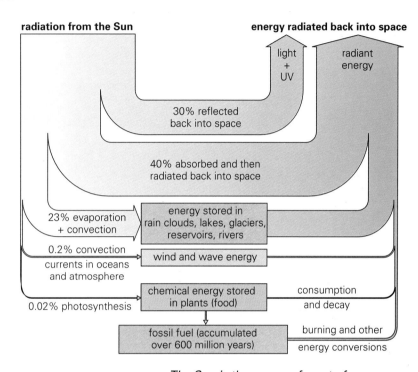

The Sun is the source of most of our energy.

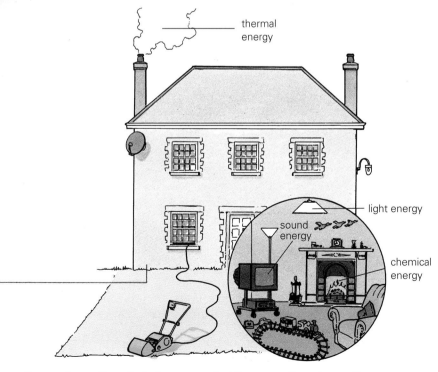

Energy is continually being converted from one form to another.

QUESTIONS

1. Give examples of three things which possess: **a** kinetic energy, **b** gravitational potential energy, **c** chemical energy, **d** mechanical energy.

2. Give two examples of energy conversions which occur naturally, and two which are caused by people.

3. What eventually happens to all the Sun's energy?

8.2 *Energy conversions*

Things as different as light bulbs, computers, cars and people have one thing in common – they all use energy.

As the energy is used it doesn't disappear, but is converted from one form to another. The total amount always remains the same.

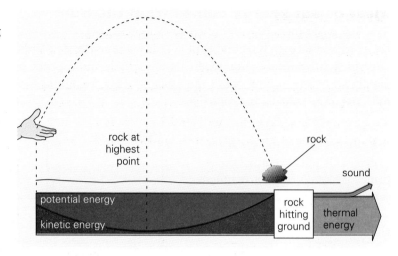

Kinetic energy is converted first to potential energy, then back to kinetic energy, then finally to sound energy and thermal energy.

Energy ends up as heat

The job of a light bulb is to produce light. In a conventional light bulb, electrical energy is used to heat a wire filament. The white-hot filament radiates infra-red and light energy. About 10% of the electrical energy consumed is converted to light, the remaining 90% being converted to thermal energy or 'heat'.

Once the light has been emitted, some may escape through a window (and travel into space), but most of it will be absorbed by the contents of the room, increasing the internal energy and making the room very slightly warmer.

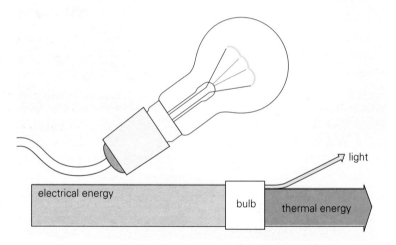

Only about 10% of the electrical energy is converted to light energy.

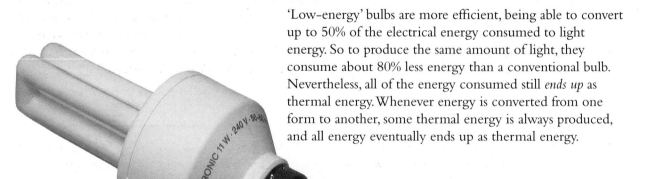

'Low-energy' bulbs are more efficient, being able to convert up to 50% of the electrical energy consumed to light energy. So to produce the same amount of light, they consume about 80% less energy than a conventional bulb. Nevertheless, all of the energy consumed still *ends up* as thermal energy. Whenever energy is converted from one form to another, some thermal energy is always produced, and all energy eventually ends up as thermal energy.

A low-energy light bulb can convert as much as 50% of the electrical energy to light energy.

Useful energy

Some energy sources, such as a battery, are concentrated and easily used. Batteries can be used in a huge variety of different devices, each of which is designed to do a different job by converting the electrical energy in a series of stages. However, as each energy conversion takes place, the energy becomes more and more 'dissipated', meaning thermal energy is produced so less of the energy stays in a form which is useful.

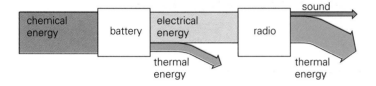

Only a tiny percentage of the energy from the batteries will end up as sound.

The thermal energy produced by people, animals and electrical equipment helps to keep a room warm.

Efficiency

The **efficiency** of a device indicates the proportion of the energy it 'wastes'. If all the energy consumed by a light bulb was converted into light energy, the bulb would be 100% efficient. The smaller the proportion of the energy converted to the required form, the lower the efficiency. Nothing is ever 100% efficient because some energy is always dissipated. You can use the following formula to work out the efficiency of a device:

$$\text{efficiency (\%)} = \frac{\text{useful energy got out (J)}}{\text{energy put in (J)}} \times 100\%$$

So, a conventional light bulb has an efficiency of about 10%.

Using dissipated energy

The heat produced by the inhabitants of a house, and their electrical equipment, contributes to the total amount of energy needed to keep the house warm in the winter. Eventually, all thermal energy escapes from a house and is dissipated more widely into the surroundings.

QUESTIONS

1. What energy changes take place in a freely swinging pendulum? Why does the pendulum eventually stop?

2. What energy conversions take place in: **a** a television, **b** a match, **c** a microphone, **d** a car engine? In each case, identify the 'useful' and the 'wasted' energy.

3. What device would you use to convert chemical energy (in one or more stages) into: **a** electrical energy, **b** kinetic energy, **c** thermal energy, **d** sound energy, **e** light energy?

4. What is the efficiency of an electric drill if 1600 J of electrical energy per second are needed to produce 1000 J of kinetic energy per second?

8.3 Transferring thermal energy

It is common sense that a hot potato will cool down and that an ice lolly will melt, but why does this happen?

Heat and temperature

The **temperature** of something tells you how hot or cold it is. It is a measure of the average kinetic energy of the molecules – the higher the temperature, the greater the average kinetic energy, that is, the faster the molecules move. Temperature is measured in **degrees celsius** (°C). Under normal conditions, water freezes at 0 °C and boils at 100 °C.

The thermal energy or internal energy of something is the *total* kinetic energy of all its molecules. **Heat** is the thermal energy that is transferred from one point to another because of a temperature difference between them – *thermal energy is transferred from hotter places to colder places*. The average kinetic energy of the molecules in the hot region falls – it cools; that of the molecules in the cold region increases – it warms up.

The transfer of thermal energy can take place by one or more of the following processes:

- **conduction**
- **convection**
- **radiation**
- **evaporation**.

Transferring energy by conduction

Conduction can take place in solids, liquids and gases. When a material is heated, the molecules which are near the source of heat gain kinetic energy, and so move faster. Each molecule passes some of its extra energy to its neighbours as it 'bumps' into them, making them move faster. The process is repeated, and energy is transferred throughout the material from the hot region to colder regions.

In a solid, neighbouring molecules are 'touching' one another. In a liquid, they are still 'touching', but each molecule 'touches' fewer others. In a gas, they aren't 'touching' at all; there are large spaces between them, and energy is only transferred from one molecule to another when they collide. Consequently, the conduction of heat is slowest in gases and fastest in solids.

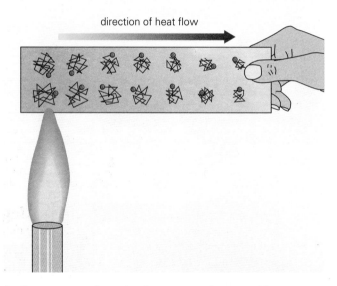

In the process of conduction, energy is passed from one molecule to the next.

Metals are particularly good **conductors** of heat, because as well as the process described above, energy is transferred from the hotter regions to the colder ones by the 'free' electrons (see 1.1), which are able to move throughout the material. They move more rapidly when heated, taking the energy with them and also transferring it by collisions.

Materials (such as gases) which are poor conductors of heat are known as **insulators**. The table shows the **thermal conductivity** of different materials. The higher the conductivity, the better the conductor. The conductivity of copper is over 10 000 times that of air. Many materials which are used for their insulating properties contain trapped air.

Material	Thermal conductivity at room temperature (W/m °C)
copper	385
aluminium	238
iron	80
Pyrex glass	1.1
brick	about 1.0 (depends on type)
rubber	0.2
air	0.03

Transferring energy by convection

When a fluid (a gas or a liquid) is heated, the molecules move faster and push each other further apart. The fluid expands and becomes less dense. The less dense fluid then rises upwards, taking its thermal energy with it. The rising fluid is replaced by cooler fluid, and a convection current is set up. Convection currents in water can be seen by placing a crystal of potassium permanganate in the beaker and then heating gently.

Thermal energy is transferred by the molecules themselves actually moving from the hot region to the cooler one. Convection cannot take place in solids because, unlike the molecules in a liquid or a gas, the molecules in a solid are unable to move from one place to another.

The arrowed lines show the direction of the convection currents.

Transferring energy by evaporation

In a liquid, the individual molecules are moving around randomly at different speeds. If a molecule which is at the surface, and moving in a direction which is away from the bulk of the liquid, has a sufficiently high speed, it is able to break free from the forces of attraction holding it to the other liquid molecules, and escape from the surface. In other, words, it will have **evaporated** and entered the gaseous state. Because it is the molecules with a higher than average kinetic energy which escape, the average kinetic energy of those remaining is reduced, and so the temperature falls and the liquid cools.

The rate of evaporation (and so the rate of cooling) depends on:

- the temperature of the liquid – the higher the temperature, the greater the rate of evaporation because more molecules will have sufficient energy to escape
- the surface area in contact with the air – the greater the surface area, the greater the rate of evaporation because more of the faster molecules will be at the surface
- the humidity of the air – a build-up of vapour above the surface will reduce the overall rate of evaporation because more vapour molecules will re-enter the liquid
- air movement above the surface – draughts prevent a build-up of vapour and so increase the rate of evaporation.

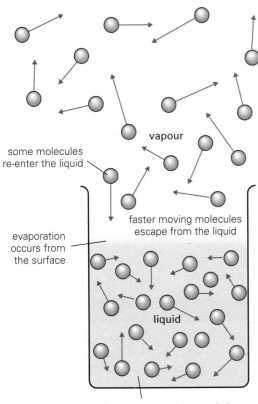

The process of evaporation.

QUESTIONS

1. Why can't convection take place in solids?

2. Why are heaters normally placed near the floor, rather than near the ceiling?

3. Does blowing over a hot drink help to cool it down? Why?

8.4 More about conduction and convection

How good a conductor?

The conductivity of different metals can be compared using either of the pieces of apparatus shown. Each of the rods should have the same diameter. The apparatus on the right is fairer, because there is more chance of all the rods being heated to the same extent.

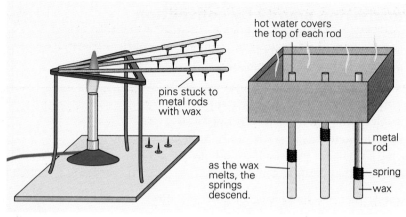

Comparing the conductivity of metals.

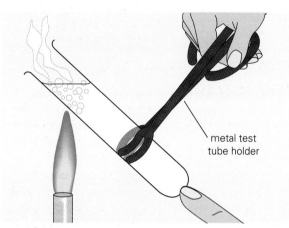

The tube can be touched at the bottom because water and glass are bad conductors of heat.

It can be shown that water is a bad conductor by touching a test tube of water at the bottom, while heating it at the top. **Wear safety spectacles** in case the test tube shatters. The water at the bottom stays cool, while the water at the top starts to boil.

If, on the other hand, a tube of cold water is heated at the bottom, the water at the top gets hot very quickly because it is heated by convection currents.

Finding out about insulators

Setting up a fair test to compare insulators is often quite difficult, because different materials usually come in different thicknesses. If, however, two identical containers holding the same amount of water at the same temperature are wrapped in identical ways with different insulators of the same thickness, then the one that cools the slowest is the one with the best insulation.

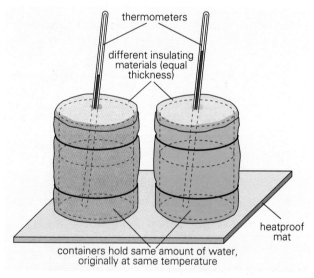

Comparing insulators.

Many materials that we use to keep us warm trap air, which is a poor conductor of heat. The air needs to be trapped in small pockets to stop it moving around and setting up convection currents. Birds keep warm in winter by ruffling up their feathers to trap more air. Several thin layers of clothing will usually trap air more effectively than one thick one. A common type of loft insulation used in houses consists of a fibreglass blanket containing trapped air. A thickness of at least 15 cm is recommended.

Filling the cavity walls with insulating foam also reduces heat loss from a house. Other gases are good insulators too. Foam polystyrene contains pockets of gas. It is used to insulate the walls of fridges and freezers, as well as to keep take-away food and drinks warm.

Convection currents in the air

Convection currents form over land which is warmer than that surrounding it. These currents of warm air are known as **thermals**, and are used by birds and gliders to lift them to a greater height.

Near the coast, the land often gets warmer than the neighbouring sea during the day. A convection current rises above the land, with cold air coming in to replace it from the sea. This creates an 'onshore' breeze. At night, the position is often reversed, with the land cooling to a lower temperature than the sea. In this situation, a breeze is set up which blows out to sea – an 'offshore' breeze.

Convection currents on a larger scale, due to uneven heating of the Earth's surface by the Sun, set up the air pressure differences which cause winds. Convection currents in the oceans also have an enormous impact on the climate.

In the absence of any prevailing wind, there is often an onshore breeze at the coast because of convection effects.

Convection currents in water-heating systems

Hot-water systems which use an immersion heater make use of convection currents during the heating process. Once the heater has been switched off, the water will start to cool. It doesn't cool evenly, however, so hot water will be obtainable for some time. This is because of the convection currents in the tank – cooler, denser water falls to the bottom and hotter water rises to the top. The hot water is run off from the top of the tank, and cold water replaces it at the bottom. As the cold, denser water enters the tank, it forms a more or less separate layer at the bottom.

Convection currents and chimneys

In homes which have gas fires or coal fires in a fireplace, the hot combustion gases pass up the chimney. It is essential for the safe operation of the fire that air is drawn into the room, although such draughts may be considered a nuisance. Some appliances are made so that they draw in the replacement air from under the floor or through the wall. Others require, by regulation, the fitting of an air vent.

Domestic hot-water system.

QUESTIONS

1. Why are saucepans often fitted with wooden handles?

2. Why does lino feel colder to bare feet than carpet?

3. What makes a duvet so effective?

4. Why are several layers of paper a good wrapping for fish and chips?

5. Why is the heating element in an electric kettle at the bottom, but the cooling element of a fridge at the top?

6. Why do rescue workers crawl close to the floor in a smoke-filled room?

8.5 Radiation

All objects radiate and absorb energy. The hotter the object the more energy it radiates. Radiant energy is one type of electromagnetic radiation (see 6.4). This means that it can travel across a vacuum. When people talk about radiant 'heat', they are talking about radiation in the infra-red region of the electromagnetic spectrum. Infra-red radiation has a lower frequency (and longer wavelength) than visible radiation. Hotter objects emit more radiation at higher frequencies than cooler ones. This is why they start to glow when heated sufficiently.

When radiant energy is absorbed by an object, its temperature rises.

What affects the amount of energy radiated?

This equipment can be used to show that the amount of energy radiated does not only depend on temperature. The cube is hollow and made of metal. Each of the vertical faces has a different finish: polished silver, white, shiny black and matt black. A thermometer (or thermocouple) is placed at the same distance from the centre of each of the four vertical faces, and the cube filled with hot water. The temperature of each thermometer rises until the rate at which they are gaining energy from the cube is equal to the rate at which they are losing it to their surroundings. Although the surface temperature of each face is the same, the temperature recorded by the four thermometers is found to be different. The highest temperature is recorded by the thermometer next to the matt black face and the lowest by the one next to the silver face. This observation can be explained by the fact that different coloured surfaces emit radiation at different rates.

In general, black surfaces radiate more energy than white ones, and matt surfaces radiate more energy than shiny ones.

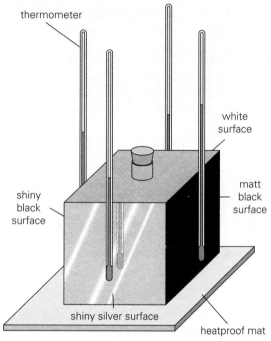

Good radiators are good absorbers

Surfaces which are good radiators of energy are also good absorbers. This apparatus gives a convincing demonstration that a matt black surface absorbs energy at a greater rate than a shiny silver one.

Both corks are held in place with wax, which melts when it gets hot enough.

108

The fire-fighter's suit reflects radiant heat.

Keeping things cool or warm

Flat roofs which are covered in bitumen felt are often covered with white stone chippings. They absorb less energy from the Sun than the black-coloured bitumen would on its own, preventing it from melting, and keeping the building cooler. They also keep the building warmer in winter, because the white surface radiates less heat. Silver-coloured roofing felts also help keep buildings cool in the summer and warm in the winter.

Fire-fighting suits are bright and shiny to protect the fire-fighters better. 'Space blankets' have a similar silvery surface, and are used by ambulance crews to reduce heat loss from accident victims.

Electrical equipment is often fitted with cooling fins. These have a large surface area and are often painted black, to increase the rate at which energy is radiated. Covering the cooling fins prevents convection from occurring, and is likely to cause the equipment to overheat.

A common way of keeping drinks hot (or cold) is to put them in a vacuum flask. These usually consist of a double-walled glass container. During manufacture, the surfaces enclosed by the double wall are coated with a thin layer of silver, and the air occupying the space is pumped out, leaving a vacuum. The stopper is usually made from an air-filled piece of plastic.

Cooling fins on an amplifier.

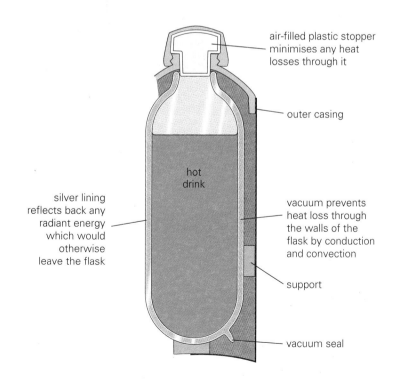

Vacuum flasks keep drinks hot.

QUESTIONS

1. Explain why the only energy reaching the Earth from the Sun is radiated energy.

2. Why is the grill at the back of a refrigerator painted black rather than white?

3. Why does dirty snow melt more quickly than clean snow?

4. How could you use less energy when boiling water in a stainless steel saucepan on a radiant ring?

8.6 Saving energy at home

How are buildings heated?

Cookers, lights and other electrical equipment, and people, all produce thermal energy and so contribute to the warming of a building as well as the heating system. During the day, buildings also absorb energy from the Sun. In new buildings, the amount of this energy that can be trapped may be increased by placing the majority of windows on the south side of the building, and by incorporating solar panels into the roof. Offices, houses and flats which are adjoined to others, may gain energy from them as well.

Some large buildings are fitted with heat pumps. They work in a similar way to a fridge. Thermal energy is extracted from a source outside the building (for example from a pond or a river which will therefore tend to get colder) and transferred to the inside.

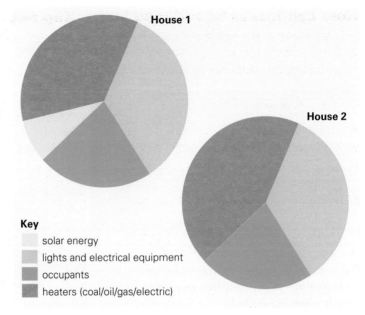

Key
- solar energy
- lights and electrical equipment
- occupants
- heaters (coal/oil/gas/electric)

Different households are heated in different ways. Which house is most energy efficient?

Where does all the energy go?

Different parts of a house lose energy at different rates (indicated by different colours in the photograph). Although, area for area, the windows lose energy at the greatest rate, in most cases the total amount that they lose is less than that from the walls or roof, because they have a much smaller total surface area.

Before permission is granted for the construction of new housing, the local authority has to be satisfied that it will meet the current requirements for energy conservation. The overall rate at which a building will lose energy can be calculated using **U-values**. The lower the U-value of a window or wall, the greater its insulating properties.

Single-glazed windows have higher U-values than double-glazed ones which, in turn, have higher U-values than walls and roofs. The higher the U-value, the higher the rate of energy transfer through a specified area under the same conditions.

U-values are measured in $W/m^2 °C$. A window that has a U-value of $4.2\ W/m^2 °C$, for example, will lose 4.2 joules of energy each second through each square metre of its surface for each °C difference in air temperature between the outside and the inside.

A house photographed with a heat-sensing camera.

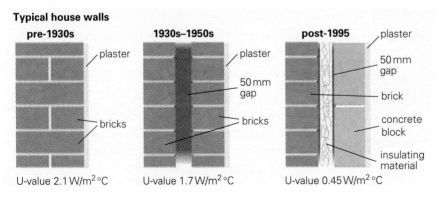

Modern house walls have lower U-values because of their different construction.

How can losses be reduced in existing housing?

When a house is at a steady temperature, it is losing energy as fast as it is gaining it. If you could reduce the rate of loss, you could therefore reduce fuel consumption, so lowering your heating bills and helping to conserve dwindling sources of energy (see 9.1).

Energy can be saved by:

- fitting loft insulation (see 8.4)
- insulating the ground/basement floor
- filling cavity walls with insulating material (by injection)
- fitting draught excluders to doors and windows
- insulating hot water tanks and pipes
- investing in low-energy light bulbs
- investing in a more energy-efficient central heating boiler (for example, a condensing boiler)
- fitting individual radiator thermostats
- having thicker curtains, curtain linings, and/or fitting shutters
- building a storm porch.
- installing double-glazing

Not all energy-saving ideas are cost-effective to carry out, however. The amount of money spent on fitting draught excluders, for example, would be recouped after only a few months, whereas it might take many years for the installation of double-glazing to 'pay for itself', by savings on heating bills.

Fitting draught excluders can make a big difference to the amount of energy used.

QUESTIONS

1. Explain why a terraced house uses less energy for heating than an equivalent detached one.

2. Explain with examples why some energy-saving ideas may be cost-effective only if they are built into a house from the start.

3. Make a list of the ways in which heating costs could be reduced in your school.

4. Why is the rate of energy transfer through a double-glazed window less than that through a single-glazed window?

5. Calculate the amount of energy lost each day through a wall of a house, if it has a U-value of 0.45 W/m² °C, an area of 25 m², and the inside temperature is 15 °C greater than the outside temperature.

8.7 Potential and kinetic energy

You can't tell an object's potential energy just by looking at it. To tell how much gravitational potential energy an object has, for example, you need to know its weight, and its height above the ground.

Gravitational potential energy is the stored energy due to work being done in raising an object against the force of gravity. The greater the weight (or mass) and/or the greater the height, the greater the gravitational potential energy.

Anything which is moving has kinetic energy. The greater the mass and/or the greater the speed, the greater the kinetic energy.

Which of the bags has the greatest potential energy?

Energy conversions in a pendulum

A simple pendulum, consisting of a bob suspended by a string, is a device which converts energy backwards and forwards between potential energy (PE) and kinetic energy (KE).

At either end of its swing, a pendulum has its maximum (gravitational) potential energy and zero kinetic energy. As the bob falls, potential energy is converted to kinetic energy, so that by the bottom of the swing, the kinetic energy is at a maximum and the potential energy is a minimum.

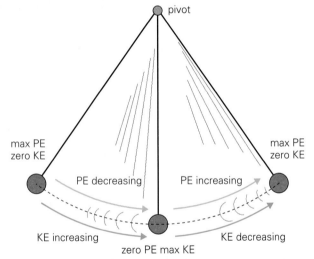

With each swing, potential energy is converted to kinetic energy, and then back to potential energy again.

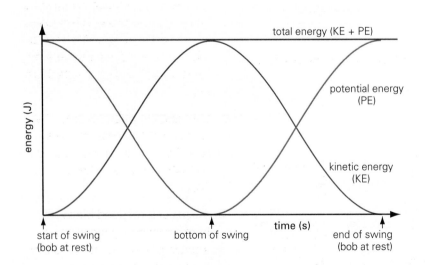

If you could have a pendulum which wasn't slowed down by air resistance and energy losses at the pivot, the conversion between potential and kinetic energy would continue indefinitely, and the total energy of the pendulum would remain unchanged. A real pendulum loses a small amount of its energy as thermal energy to the surroundings with each swing, and so eventually it comes to a stop.

The energy of a pendulum is stored as kinetic and potential energy.

Calculating gravitational potential energy

The amount of gravitational potential energy an object has can be calculated using the equation:

gravitational potential energy (J) = weight (N) × height (m)

and since weight = $m \times g$ (see 4.1) this can be rewritten as:

gravitational potential energy (J) = mass (kg) × gravitational field strength (N/kg) × height (m)

or, in symbols: PE = $m \times g \times h$ or   PE = mgh

Sometimes, it is more useful to be able to calculate the *change* in potential energy of an object when it moves.

change in potential energy (J) = weight (N) × change in vertical height (m)

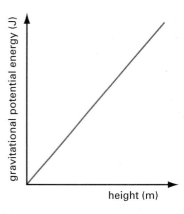

Gravitational potential energy increases uniformly with height.

Calculating kinetic energy

The amount of kinetic energy an object has can be calculated using the equation:

kinetic energy (J) = $\frac{1}{2}$ × mass (kg) × (speed)^2 [(m/s)^2]

or, in symbols: KE = $\frac{1}{2} \times m \times v^2$ or   KE = $\frac{1}{2}mv^2$

where v is the object's speed.

Doubling an object's mass (but keeping the speed the same) will double its kinetic energy. Doubling its speed (but keeping the mass the same) will quadruple the kinetic energy. A car travelling at 70 mph will have roughly twice the kinetic energy of a similar car travelling at 50 mph. (and will need twice as much force to stop it in the same distance).

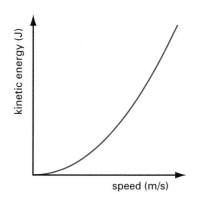

Kinetic energy increases with the square of the speed.

The speed of falling objects

The potential energy of a falling object is converted to kinetic energy as it falls. If the object falls from rest, then the change in potential energy (weight × change in height) will equal the gain in kinetic energy:

$$m g h = \frac{1}{2} m v^2$$

where h is the distance fallen and v is the speed of the object when it has fallen this distance.

When the object is about to hit the ground, all its potential energy will have been converted to kinetic energy. When it actually hits the ground, the kinetic energy will be converted into thermal energy and sound energy. Provided the distance fallen is not too large and the mass not too small, the effect of air resistance can be ignored (as it has been here).

As an object falls, its potential energy is converted to kinetic energy.

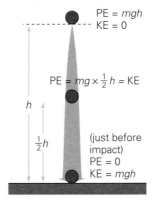

QUESTIONS

Take g = 9.8 N/kg.

1. Calculate the potential energy of a bag of mass 2.5 kg which is resting on a chair 0.5 m high. What will be the change in its potential energy if it is then placed on a table 0.9 m high?

2. Calculate the kinetic energy of a car of mass 900 kg travelling at a speed of 20 m/s. At what speed will its kinetic energy have doubled?

3. An object is dropped from rest from a window. Ignoring any air resistance, calculate its speed by the time it has fallen 10 m.

8.8 Work and power

What is work?

Some words used by scientists can be confusing because their scientific meaning is different from their everyday meaning. You have already come across this with the word 'weight'. The word 'work' also has a special scientific meaning.

In science, we say that work has been done on something when energy is transferred to it as a result of some sort of movement caused by a force. When you push a book to one side, for example, or stretch a spring, you are doing work on the book or the spring. Energy is needed to do the work and it is converted from one form to another in the process.

Working out how much work a force has done

The amount of work done can be calculated by using the formula:

work done (J) = force (N) × distance moved in the direction of force (m)

This is often written simply as:

work done (J) = force (N) × distance (m)

or, in symbols:

$W = F \times d$

but you must be careful to use the correct distance – the distance moved in the direction of the force.

Work is measured in joules (J).

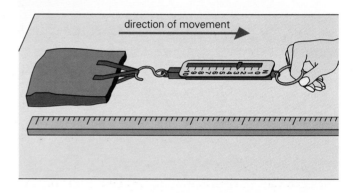

If the bag is pulled a distance of 0.5 m, the work done will be 0.5 m × 3 N = 1.5 J.

Looking at energy changes

It is often useful to think about the energy changes which occur as work is being done. When work is done on something there is a transfer of energy to it. A bag being carried upstairs, for example, gains potential energy as it rises. *The amount of work done is equal to the amount of energy transferred.*

So the work done on the bag is equal to its gain in potential energy. The lifting force (F) on the bag is equal to its weight. The distance moved *in the direction of the force* is the change in vertical height. So

change in potential energy (J)

= work done (J)

= weight (N) × change in height (m)

Sometimes, when work is done, the energy is transferred in more than one way. When the bag is pulled along the bench or the ground, for example, it will gain kinetic energy but thermal energy will also be produced because of frictional resistance.

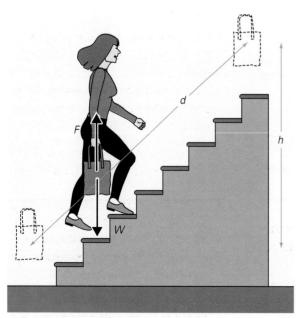

Although the bag will have moved d metres, it will only have moved h metres in an upward direction (the direction of the force F).

Power

Power is a measure of how fast energy is being transferred or converted – *or* how fast work is being done. Power is measured in watts (W). The unit is named after James Watt, a Scottish engineer and inventor who made a vital contribution to the practical development of steam engines. One watt is equivalent to one joule per second (J/s) – a 100 W light bulb converts 100 J of electrical energy (into radiant and light energy) each second.

$$\text{power (W)} = \frac{\text{energy transferred (J)}}{\text{time taken (s)}} \quad \text{or:} \quad \text{power (W)} = \frac{\text{work done (J)}}{\text{time taken (s)}} = \frac{\text{force (N)} \times \text{distance (m)}}{\text{time taken (s)}}$$

The 'power output' of a device tells you how much energy is converted into the required form each second. It is always less than its 'power input' (or 'power consumption') – the rate at which energy is supplied to the device – because of dissipated energy (see 8.2).

The power consumption/output of some devices is quite large, so may be measured in kilowatts (kW) or megawatts (MW).

1 kW = 1 000 W
1 MW = 1 000 000 W

The table gives some idea of the magnitude of the power output (O) or power consumption (C) of some devices

Approximate power output/consumption (W)	
Ince B power station (oil-fired) (O)	1 000 000 000
Bryn Titli wind farm (O)	10 000 000
lorry (C)	100 000
small gas boiler (O)	10 000
one-bar electric fire (C)	1000
light bulb (C)	100
torch bulb (C)	10
passive infra-red (PIR) detector (C)	0.1
calculator (C)	0.001

The maximum power output of engines is often expressed in 'horsepower' rather than in watts or kilowatts. At one time, the horsepower was only a rough unit of measure. Now the value of one horsepower is standardised, with a fixed value of 746 W.

Efficiency

The efficiency of a device (see 8.2) can be expressed as:

$$\text{efficiency (\%)} = \frac{\text{useful energy output (J)}}{\text{work done or energy input (J)}} \times 100\%$$

$$\text{or} \quad \frac{\text{useful power output (W)}}{\text{power input (W)}} \times 100\%$$

The efficiency of many machines can be increased by ensuring that they are properly lubricated, so that as little energy as possible is 'wasted' in doing unnecessary work against friction.

There is a practical limit to the efficiency of some machines. Consider a simple pulley lifting a load. If the pulley was 100% efficient, then all the energy input would be converted into the potential energy of the load. However, in raising the load, the lower block of the pulley also has to be raised. The gain in potential energy of this block is 'wasted' energy.

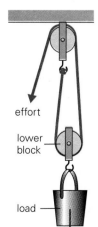

*The **useful** work done by the effort is equal to the gain in potential energy of the load.*

QUESTIONS

1. A box was pushed a distance of 1.8 m across a room, by a horizontal force of 20 N. How much work was done?

2. Calculate the work done, when a bag of mass 2.5 kg (weight 25 N) is carried up a flight of stairs, if the horizontal distance travelled is 3.5 m and the vertical distance travelled is 2.5 m.

3. a Calculate how much energy a 100 W light bulb will consume if it is left on for 3 hours.
   b A low-energy bulb which provides the same illumination has a power rating of 20 W. How long could this be left on, for the same energy consumption as the ordinary bulb in **a**?

4. Calculate the efficiency of a pulley, if 2500 J of energy are needed to raise a load of 200 N by 10 m.

CHAPTER 9: ENERGY SUPPLIES

9.1 Using energy

Available energy sources

Most of the energy sources available for us to exploit are stores of energy which originally came from the Sun (see 8.1).

	Primary source	Form of energy	Means of conversion to electricity	Other energy uses
From the Earth	hot rocks	geothermal	steam turbine	used directly in local heating schemes
	tides	gravitational potential	water turbine in a tidal barrage	
	uranium	nuclear	steam turbine	
From the Sun today	solar energy	electromagnetic radiation	solar furnace/steam turbine solar cell	heating by using solar panels
	wind	mechanical	wind turbine (windmill)	transport (sailing boat)
	rain water	gravitational potential	water turbine in a hydroelectric plant	
	ocean energy — wave	mechanical	water turbine	
	ocean energy — thermal	thermal	Ocean Thermal Energy Conversion (OTEC) plant	
	plants	biomass (chemical)	steam turbine	food; biofuels for heating, transport (by combustion)
From the Sun going back millions of years	peat, coal, oil, gas	chemical	steam turbine	transport, heating

Some energy sources, such as wind or wave energy, are rapidly renewed on a daily basis and are described as **renewable sources**. Others, such as fossil fuels, are renewed only over a much longer time scale and are being used faster than they are being replaced. They are known as **non-renewable sources**. Wood is considered as a renewable source since, with good management of forests, a continuous supply is possible. Different energy sources need to be exploited in different ways.

Primary sources of energy and their uses.

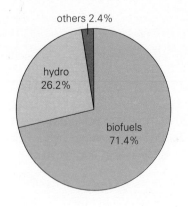

Renewable energy use in the UK in 1994. (Source: DTI)

What sources of energy do we use?

Most of the energy that we consume in the UK comes from three fossil fuels: gas, oil and coal. They are used for heating, transport and generating electricity. In 1994, they provided about 90% of the energy actively used for these purposes. Fossil fuels are non-renewable sources of energy, which means that they will eventually run out. Different groups of energy specialists are continually making forecasts about when this will be. Although they do not agree precisely, it is generally agreed that the world's supply of gas and oil will be exhausted during the first part of the next century. Coal is expected to last about another 250 years. As fossil fuels start to run out, the remaining supplies will be more and more difficult to extract and will, inevitably, become more expensive. It is clearly essential that these fuels are conserved as far as possible by using less energy, and that other sources of energy are exploited.

In 1994, just over 2% of our electricity was produced from renewable sources. The amount of electricity obtained from renewable sources has increased each year since the Electricity Act was passed in 1989. The Act allows the government to require the electricity suppliers to obtain a certain amount of their electricity from renewable sources of energy. Renewable sources, especially bio-fuels from plants and organic matter, are also being used more and more for heating and transport.

A major source of energy not provided by the Sun is nuclear fuel. This is a non-renewable source, but supplies of the main fuels uranium and plutonium are still plentiful. These are used in nuclear power stations for electricity generation. The nuclear power industry in the UK expanded up until the mid-1990s, by which time it was providing about a quarter of the electricity supply. The French nuclear industry, by contrast, supplies the majority of France's electricity.

Making fuels more useful

One of the most useful and convenient sources of energy is electricity. Electricity has to be produced from fuels or other 'primary sources' of energy – it is a 'secondary source'. The fuels provided by the Sun, renewable sources such as wind energy (also powered by the Sun), nuclear fuels and geothermal energy are all called primary sources.

Producing electricity from a primary source can waste a lot of energy – as much as two-thirds in some fossil fuel power stations. In the UK in 1992, only 65% of the energy available from primary fuels was actually used by the final user: 22% of it was lost in the conversion to electricity.

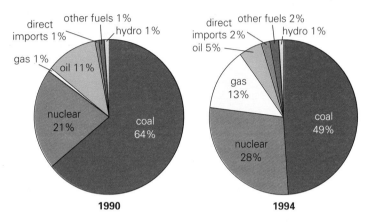

How we get our electricity in the UK. (Source: DTI)

Using energy has an environmental cost

Whether we like it or not, because we use energy, we contribute to the pollution of our environment. Simply getting the energy to us affects the environment, as well as its use. Some forms of energy cause less environmental damage than others, but all have some environmental cost.

Oil spills damage the atmosphere as well as the marine environment.

QUESTIONS

1. Why is electricity a very convenient source of energy to use?

2. What environmental problems are there in your local area due to energy production, distribution or use?

3a. Describe how the way in which the UK obtained its electricity changed between 1990 and 1994.
  b. What implications might this have for the future?

9.2 Power stations

However or wherever our mains electricity is generated, all the generating systems have at least two components in common: a turbine and a generator (see 3.4 and 3.6). The generator is turned by the turbine, which is itself turned by a high-pressure flow of fluid – air (wind), combustion gases, steam or water – past its blades.

The blades of a steam turbine look very similar to those of a jet engine.

Most of the electricity in the UK is generated using steam-driven turbines. The steam is produced by heating water with the energy obtained from:

- burning fossil fuels
- nuclear reactions in a nuclear reactor
- burning biomass such as straw or rubbish.

Water is used to turn the turbines in hydroelectric, tidal and wave generators. Commercial power stations vary enormously in size, from for example large coal-fired stations capable of producing around 2000 MW down to much smaller plants using wind or wave energy whose output is, perhaps, a thousand or more times less.

Burning fossil fuels

When fossil fuels are burnt, a variety of combustion products are emitted into the atmosphere, the amounts depending on the type of fuel, and its state of purity:

- dust particles – these are removed from power station emissions by electrostatic precipitation (see 1.2)
- sulphur dioxide, which causes acid rain when it combines with water in the atmosphere – this can be removed from power station emissions by passing the flue gases through a slurry of crushed limestone, but this is expensive
- carbon dioxide, a 'greenhouse gas', is a major contributor to global warming since it absorbs infra-red radiation from the Earth
- nitrogen oxides and other pollutants – the amount of these in vehicle exhausts can be reduced by fitting catalytic converters.

Gas – a cleaner fuel?

For the same amount of energy, natural gas produces less carbon dioxide than either oil or coal when burnt under comparable conditions, and also less of the other undesirable products.

Another advantage of gas is that it can be burnt more efficiently using a 'combined cycle gas turbine' (CCGT). The combustion gases turn one turbine, while the steam from the boilers turns another. CCGT power stations cost less to build and can be up to 30% more efficient than conventional coal-fired power stations.

Unfortunately, gas has one major disadvantage; the reserves of natural gas will run out long before the reserves of coal.

The CCGT power station at Roosecote, Cumbria.

When the nuclear reactor at Chernobyl blew up, the whole of Europe was affected by the radioactive fallout.

Nuclear power stations

In nuclear power stations, uranium or plutonium undergo radioactive decay in nuclear reactors (see 10.7). These nuclear reactions produce thermal energy which is used to produce the steam to drive the turbines. The reactions are carefully controlled to prevent a 'runaway' effect and possible explosion.

This is arguably a much cleaner way to produce electricity than burning fossil fuels, but there are major disadvantages:

- the cost of building nuclear power stations is high, as is the cost of decommissioning them at the end of their useful life
- people have fears of radioactive substances – they are concerned about the safe transportation of fuels and disposal of waste materials
- there is the risk of nuclear accidents.

Siting power stations

Power stations are built in particular locations for particular reasons. Steam turbines need a large supply of water to condense the used steam after it has passed through the turbines. Power stations are therefore often built on the coast or on the banks of large rivers. They take in cold water and return it at a higher temperature. Others need to be built at or near the site of the energy source, either because the energy source can't be transported – for example, wind or wave energy – or because the fuel is expensive to move over large distances, for example coal.

There are many environmental or social as well as economic considerations to be taken into account. For example, steam-driven turbines built away from a natural water supply require huge unsightly cooling towers to recycle their water. Nuclear power stations are built away from highly populated areas, in case of radioactive emissions through accidents. A tidal barrage cannot be built where it would disrupt major shipping routes.

The River Thames at Battersea.

QUESTIONS

1. Make a list of the factors that you would want to be taken into account if there was a proposal to build a power station near you.

2. The River Thames used to have several coal-fired power stations along its banks in London. What effect would they have on the river? How do you think the coal was delivered?

3. Make a table showing the advantages and disadvantages of generating electricity from coal, gas and nuclear fuels.

9.3 Using water to generate electricity

Hydroelectric power

Hydroelectric power stations collect and store water from a large area in a reservoir or series of reservoirs. When electricity is needed, the water is allowed to fall under gravity through a pipe connected to a turbine. The turbine drives the generator which produces the electricity. Gravitational potential energy is converted to electrical energy.

Unlike a coal-fired power station, which can't easily be switched on and off, the hydroelectric power station at Rheidol, for example, can be switched on and off in under four minutes! Hydroelectric power stations are therefore used to 'top-up' the National Grid at times of peak demand. The Rheidol power station operates for an average of four hours per day, and is capable of producing 49 MW of power.

Sometimes more power is supplied to the National Grid than is needed to meet demand. This energy can be stored at a pumped hydroelectric station such as the one at Dinorwig in north Wales. The surplus electricity is used to pump water from a low reservoir to one at a higher level. The pumped water gains potential energy, which can then be converted back to electrical energy at times of peak demand.

Hydroelectric power stations need to be built in hilly areas with a reliably high rainfall. These are often areas of great natural beauty.

Tidal power

At present, there are no tidal power stations operating in the UK. The one shown here is located at the mouth of the river Rance in northern France. It works in a similar way to a hydroelectric power station, and doubles as a bridge. When the tide comes in, the storage reservoir is filled with water. As the tide goes out, the stored water can be used to operate the turbines. High tides occur on average every 12 hours 25 minutes, so the times at which it is able to generate vary on a daily basis, but since tides are reliable, so is the daily supply.

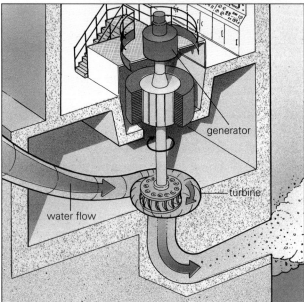

A cross-section through one of the Rheidol generators.

In order to be viable, a tidal power station needs to be built across the mouth of a river where there is a large tidal range (a large difference in water level between high and low tides). The maximum tidal range at the mouth of the Rance is about 13.5 m. At most places it is very much less than this. It has been suggested that a tidal power station could be built across the Severn estuary. If it was built, it would have a larger output than the one across the Rance. It would probably have a maximum output of over 5000 MW and be capable of supplying over 3% of the UK's electricity. It would have a major environmental impact on the plants and animals in the Severn estuary, however, and would affect shipping.

The tidal power station on the Rance has a peak output of 240 MW.

Wave power

The energy from waves can be converted into electricity in a variety of ways. The oscillations of the waves have to be changed by some kind of wave-energy converter into a rotary motion to drive a turbine. In theory there is enough wave energy around the coastline of Britain to provide us with all our electricity. Wave generators are still at an early stage of development, however, and only a tiny fraction of our electricity is produced in this way because of practical difficulties.

In 1991, electricity generated from wave power was fed into the National Grid for the first time, from a small (75 kW) experimental on-shore plant on Islay in the Inner Hebrides. In the summer of 1995, the world's first commercial wave generator was launched in Scotland. Named Osprey 1, it was designed to produce enough electricity for 2000 homes (a total of 3.5 MW, of which 1.5 MW was to come from a wind turbine bolted on top). Unfortunately, within weeks, it had been holed by the sea and become inoperable.

The ill-fated Osprey 1 wave generator.

Extracting thermal energy from the sea

The surface waters of the oceans are up to 20°C warmer than the water at a depth of several hundred metres. An Ocean Thermal Energy Conversion (OTEC) plant aims to exploit this temperature difference to extract thermal energy from the warmer water at the surface.

QUESTIONS

1. In which parts of the UK would you expect to find a hydroelectric plant?

2. In what ways might a tidal power station affect the environment?

3. Why are hydroelectric power stations important?

4. How much potential energy is stored in a reservoir, if it contains 180×10^6 cubic metres of water, at an average height of 20 m above the turbines? (Assume one cubic metre of water has a weight of 10 000 N.)

9.4 Energy from wind, hot rocks and waste

Wind energy

Windmills convert the kinetic energy of the wind into a more useful form of energy. Modern windmills (or wind turbines) produce electricity – the shaft of the rotating blades is connected directly to a generator. They differ considerably from traditional windmills in their size, appearance and power output. A wind turbine with a blade span of 50 m can generate 1 MW of electricity. It would take about 50 000 traditional windmills to produce the same amount of power!

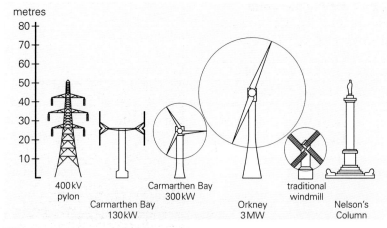

Wind turbines come in different shapes and sizes. Some have blades which rotate in a horizontal rather than a vertical circle.

In general, the windiest places tend to be on the top of hills or near the coast – places which people value for their natural beauty. Wind turbines are often built in groups of 20 to 30 in 'wind farms'. Some people think they add to the beauty of the landscape, but others think differently and are prepared to say so. Apart from the aesthetic considerations, other factors that need to be taken into account include the effect on bird life, the effect on local TV reception, and the effect of blade noise on people living nearby.

It has been estimated that enough electricity could be generated from wind turbines built on land not set aside for its natural beauty to provide between 20% and 50% of the UK's current consumption.

A wind farm.

The one big disadvantage of wind power is that winds are unpredictable and unreliable. The wind has to have a certain minimum speed before the blades will start to turn, and the power then available depends on the cube of the wind speed. Because of this, wind power is most likely to be used to supplement conventional means of producing electricity rather than replace them.

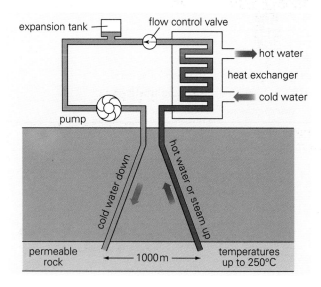

Geothermal energy

The deeper you go underground, the warmer the rocks get. In some places, however, the warmer rocks are nearer to the surface than in others. In parts of Cornwall, for instance, because of the presence of radioactive material in the rock, temperatures of 200 to 300 °C exist just a few kilometres below the surface. By drilling down to them, and using pumped water, some of the thermal energy could be extracted. The most efficient use of the hot water would be to pipe it to institutions such as schools and hospitals for use in their heating systems (a district heating scheme – see 9.6). At present, geothermal energy is not exploited in the UK.

Extracting energy from hot rocks.

In some parts of the world, however – for example, Iceland – the rocks near the Earth's surface are so hot that the water in streams and lakes can be too hot to touch. Geothermal energy makes an important contribution to Iceland's total energy needs.

At this geothermal power station in Iceland the turbines are driven by steam heated by hot rocks.

Energy from waste

Each day, thousands of tonnes of rubbish are produced, and have to be disposed of. Traditionally, much of this waste has been disposed of in landfill sites, rather than being recycled. Not only does this waste valuable resources which could be recycled, but also, as organic material decomposes, it produces methane – a greenhouse gas (see 9.5) – which can seep from the ground and escape into the atmosphere.

Local councils have been ordered by the government to recycle 20% of their waste by the year 2000. One result of this has been the building of hi-tech plants such as the one which opened in 1994 in Lewisham, south London. It is capable of burning 420 000 tonnes of rubbish each year. The combustion process is strictly controlled to prevent the formation of harmful chemicals and to ensure that the plant operates within the limits set by the EU on emissions. As well as generating up to 32 MW of electricity, it is capable of producing hot water for a district heating scheme.

Some power stations are designed to burn specific types of waste. For example, a 28 MW power station near Wolverhampton uses old tyres as its fuel, and a 14 MW one in Suffolk burns chicken litter.

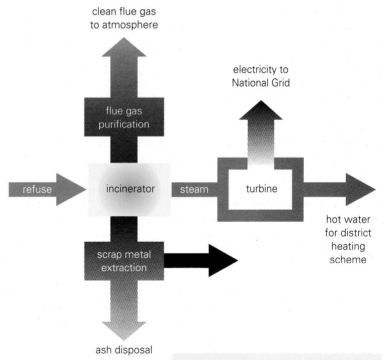

The recycling of energy and materials from waste.

QUESTIONS

1. In which parts of the UK might you find wind turbines?

2. Explain why wind energy is not a reliable source for continuous electricity production.

3. Why is it better to recycle waste than bury it in landfill sites?

9.5 Using solar energy

Solar furnaces
Solar furnaces are designed to collect and focus the Sun's energy. This is achieved by using as many as 20 000 computer-controlled curved mirrors. The energy is used to drive a steam turbine and generate electricity. The world's first large-scale solar furnace started operating in 1981 in Italy, generating 1 MW. In order to be economically viable, solar furnaces are built in places which have only a small amount of cloud cover throughout the year, and usually nearer to the Earth's equator than its poles.

The solar furnace in Font-Romeu, France.

Solar cells
Solar cells are electronic devices which convert the Sun's energy directly into electrical energy. The best have an efficiency of about 17%, which means that for every 100 J of energy falling on them, they will produce about 17 J of electrical energy. Each square metre of solar cells can therefore produce a maximum of about 35 W of power. Solar cells are commonly used in low-power devices such as calculators. They are also used for small-scale electricity generation in remote parts of the world where there is no other supply available, and to power equipment in satellites (see 3.3).

Despite their high cost and low efficiency, solar cells can be used as they have been at Serre in Italy, to generate significant quantities of power. The Serre plant has a maximum output of 3 MW.

Active solar heating (ASH)
Solar panels can be used to heat water in the home or in swimming pools. They work by absorbing the radiant energy from the Sun. When it is cloudy, there is still energy available to be collected, but not as much. Solar panels are most often seen installed on the south-facing slope of a roof. In other locations, they can be made more efficient by being computer-controlled so that they are continually turned to face the Sun.

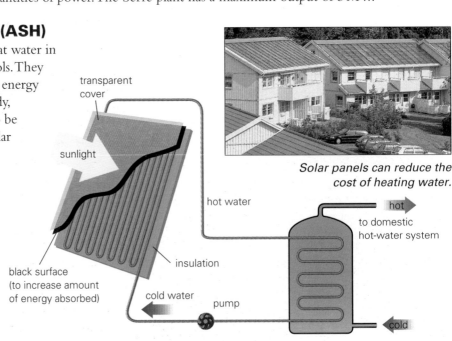

Solar panels can reduce the cost of heating water.

Making use of the greenhouse effect

Greenhouses get warm because glass will only allow certain frequencies of electromagnetic radiation to pass through it. During the day, visible and short-wavelength infra-red radiation pass through the glass into the greenhouse, where they are absorbed, making the inside of the greenhouse warmer. You have already seen (8.5) that all objects radiate thermal energy. The energy radiated by the contents of the greenhouse is in the longer-wavelength infra-red region of the electromagnetic spectrum. Glass will not transmit this, so the energy is 'trapped' in the greenhouse.

This 'greenhouse effect' can be made use of by architects when designing houses, to trap solar energy and reduce heating costs. Groups of 'solar houses' are arranged so that the sunlight falling on each house is obstructed as little as possible.

The greenhouse effect is thought to be responsible for global warming. The 'greenhouse gases' in the atmosphere – for example, carbon dioxide, methane and CFCs – are equivalent to the glass in the greenhouse. They allow short-wavelength infra-red, visible and ultra-violet light from the Sun to pass through them, but trap the long-wavelength infra-red radiation being radiated by the Earth.

Energy from plants

Plants trap the Sun's energy, and convert it into chemical energy by the process of **photosynthesis**. Straw, a by-product of wheat production, can be used as a fuel in a suitably designed power station. The best place for such a power station would be in an area where large quantities of wheat are grown.

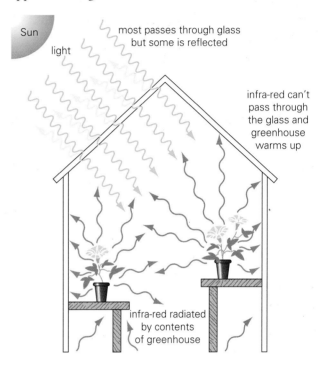

The greenhouse effect.

Some plants, such as sugar cane, are particularly efficient at trapping and storing the Sun's energy. The sugars can be converted to ethanol, a liquid biofuel, by the process of **fermentation**. In this process, yeast breaks down the sugar to produce ethanol (alcohol), carbon dioxide and water. One use of ethanol is to mix it with petrol to make a fuel called 'gasohol' for use in cars. This reduces the polluting emissions such as sulphur. Vehicles can also run on pure ethanol, but engine modifications are necessary.

Sugar cane traps the Sun's energy efficiently.

QUESTIONS

1. Why are solar cells so useful in some places?
2. How would you design a house to maximise the amount of energy it could gain from the Sun and, at the same time, keep heat losses to a minimum?
3. Describe how a solar panel collects energy.

9.6 Energy and money

The cost of generating electricity at any particular power station depends on a number of factors, which include:

- the cost of the fuel
- the capital costs (the site, the generating equipment, the buildings, connection to the National Grid)
- the efficiency of energy conversion
- the cost of personnel
- the cost of maintenance
- the cost of environmental and safety measures
- the expected life-span of the installation
- the decommissioning costs.

Nuclear power stations, for example, are likely to have huge and not easily predictable decommissioning costs at the end of their working lives. Although the wind and sunlight are free, it still costs money to generate electricity from them. Harnessing significant useful power from renewable sources requires, kilowatt for kilowatt, a relatively high capital investment.

Different groups produce different figures about the economic, as well as the environmental, costs of generating electricity at different types of installation. The financial calculations and arguments are complex, so, not surprisingly, different experts don't always agree about the overall costs involved!

Increasing the efficiency of power stations

The efficiency of fossil-fuel power stations has improved markedly over the years. In 1994/95, the average efficiency of those using conventional steam turbines in the UK was just over 36% (which means on average 640 J out of every 1000 J of energy produced by the fuels is wasted). The efficiency of combined cycle gas turbine (CCGT) power stations (see 9.2) is greater, averaging 46% in 1994/95, with those still under construction having design efficiencies of up to 55%.

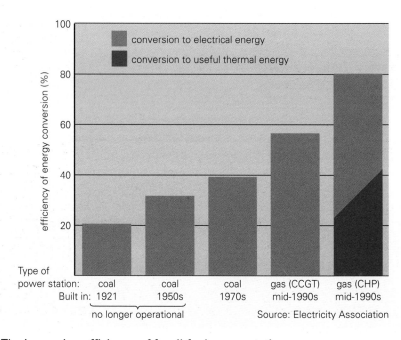

The increasing efficiency of fossil-fuel power stations.

The overall efficiency of new power stations is dramatically increased if they are built as 'combined heat and power' (CHP) installations. These produce not only electricity, but also thermal energy in the form of piped hot water, suitable for use in the heating systems of large energy users such as schools, hospitals and factories. The percentage of energy converted into electrical energy is less than in a conventional power station, but the overall conversion into electrical energy and useful thermal energy can be as high as 80%. In other words, they are twice as efficient as many conventional power stations. They need to be built in towns, near the users of the hot water and, in practice, this means that they tend to be of limited size and generating capacity.

Matching supply to demand

On a typical day, demand for electricity can be predicted very accurately.

The minimum demand (the 'base load') in the UK is largely met from the nuclear power stations (which are difficult to switch on and off) and from the cheapest sources such as CCGT power stations and French nuclear power stations. Supply is increased as demand rises. The order in which different power stations are called into action is influenced by a variety of commercial factors. Peak demand is met by installations which can be started up very quickly, for example, hydroelectric stations.

Electricity is sold more cheaply in the early hours of the morning to some consumers (see 2.4), to encourage the use of electricity in what would otherwise be a slack period.

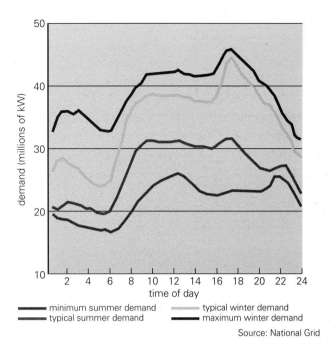

Daily electricity demand during 1994/95.

The output of power stations is carefully controlled to match supply to demand.

If supply and demand get out of balance, surplus electrical energy can be stored in the form of potential energy in the National Grid's pumped storage hydroelectric power stations at Dinorwig and Ffestiniog in Wales and Scottish Power's pumped storage plant in Cruachan. Although these plants reduce the amount of wastage that would otherwise occur, the process of converting electrical energy to potential energy, and back to electrical energy again still results in an overall energy loss.

QUESTIONS

1a Describe how the demand for electricity varies throughout the day on a typical day in the winter.
b What do you think causes the three peaks in demand on a typical day in the winter?
c How does the graph differ for a typical day in summer?

2a For what reason, other than cost, is it important to make fossil-fuel power stations as efficient as possible?
b By what factor did the efficiency of newly built coal-fired power stations increase between 1921 and the 1970s? (Refer to the bar chart opposite.)

SECTION D: QUESTIONS

For these questions take the Earth's gravitational field strength g as 9.8 N/kg.

1a What factors would need to be taken into account when choosing where to locate each of the following?
  i A tidal barrage
  ii A coal-fired power station
  iii A combined cycle gas turbine (CCGT) power station
  iv A nuclear power station
  v A combined heat and power plant (CHP)
  vi A hydroelectric power station
  vii A wind turbine
b In each case, state two possible environmental objections that might be raised.
c Why are CCGT plants more efficient than an ordinary gas-fired power station?

2 Explain why almost all of the energy we use came originally from the Sun.

3a Explain why a bungalow is likely to be more expensive to heat than a similarly constructed two-storey house of the same volume.
b State two reasons why it is probably more sensible to fit draught excluders to the doors and windows of a house than to change single-glazed windows for double-glazed ones.
c Why is a room likely to need less heating, if more electrical equipment is installed?
d How would the number of occupants affect the heating requirements of a room?

4 Explain why money can be saved if the surface of the water in a swimming pool is covered with a plastic sheet at night.

5 Why is less energy needed when cooking in copper or aluminium saucepans than iron ones of the same dimensions?

6 A plumber is installing a radiator, the height of which has to be 0.5 m, and the length has to be 1.0 m. She has the choice of single or double panel, with or without fins (diagrams i–v).

a Other things being equal, which radiator will emit the greatest amount of heat and why?
b State two things, other than changing the radiator, which could be done to increase the heat output of the radiator.

7a A ball is dropped from a window 4 m above the ground. Assuming there is no air resistance, what will its speed be just before it strikes the ground?
b From what height would it have to be dropped to hit the ground at twice this speed?

8 At what vertical speed must a ball leave the thrower's hand, 1.5 m above the ground, if it is to rise vertically to a height of 6.5 m?

9 The turbines of a hydroelectric power station are 50 m below the level of the surface of the reservoir. Assuming an efficiency of 50%, what mass of water would have to pass through the turbines each second to produce a power output of 500 kW?

10a Name two gases which are released into the atmosphere when fossil fuels are burnt in power stations.
b Explain how the release of each of these gases may eventually damage the environment.

11 Name a device which is designed to convert:
a sound energy into electrical energy
b electrical energy into thermal energy
c kinetic energy into thermal energy
d elastic potential energy into kinetic energy
e kinetic energy into sound energy
f electrical energy into sound energy
g chemical energy into electrical energy.

12 An electric light bulb is marked 230 V, 150 W.
a How much energy is converted by the bulb each second?
b What are the two main forms of energy that the electrical energy is converted to?
c If the light bulb has an efficiency of 10%, how many joules of energy are converted into light energy each second?

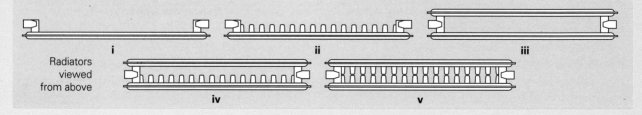

Radiators viewed from above
i   ii   iii
iv   v

13 What are the advantages of living in a house which has south-facing windows with shutters?

14 A car of mass 1200 kg travelling at 20 m/s along a level road is slowed to 10 m/s by applying the brakes. What proportion of its kinetic energy is changed into thermal energy?

15 A power of 3.08 kW is needed to tow a trailer at a steady speed of 14 m/s. What is the size of the force opposing the motion of the trailer?

16 What is the output power of a pump, if it can raise 450 kg of water a vertical distance of 90 m each minute?

17 A boy, who has a mass 55 kg, ran from the bottom to the top of the stairs as shown in 6 seconds.

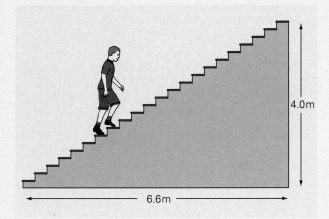

 a How much gravitational potential energy did he gain?
 b How much power did he develop in his legs?
 c How much chemical energy was converted in his legs, if the efficiency of the muscles is 20%?

18 List six ways in which heat can be lost from a house. State one way in which each of these energy losses could be reduced.

19 The diagram shows a device commonly used to measure relative humidity (how much water vapour the air contains). The drier the air, the greater the difference between the temperatures shown by the two thermometers. Use simple kinetic theory to explain why this should be.

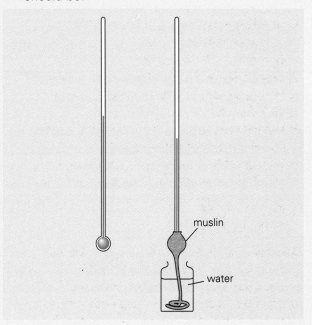

20 The graph shows how the power output of a wind generator varies with wind speed.
 a At what wind speed do the blades start to turn?
 b At what wind speed is the generator shut down to prevent damage to the blades?
 c What is the maximum power output?

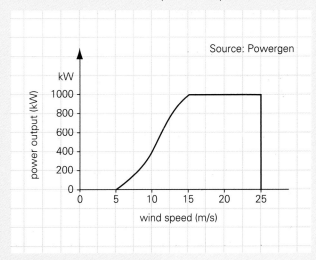

CHAPTER 10: RADIOACTIVITY

10.1 *What is radioactivity?*

The French scientist Henri Becquerel had been working with a uranium compound in 1896, when he discovered by accident that some unexposed photographic material had been affected. Not only had some form of radiation come from the uranium compound, but the radiation had passed through the light-proof wrappings of the photographic material as well.

Marie Curie, a Polish scientist working in France, pioneered further investigations into this effect, which she called **radioactivity**.

Marie Curie, 1867–1934.

Where does radiation come from?

In chemical reactions, the atoms stay the same, but simply combine in different ways. The energy changes that occur, come from changes in the chemical bonds between the atoms. Radioactivity is a result of nuclear reactions – a rearrangement of the subatomic particles takes place within an unstable nucleus of an atom, causing energy changes much greater than those of chemical reactions. **Nuclear radiation** is emitted and often one element changes to another. This is called **radioactive decay**.

The structure of atoms

All atoms are made up of three types of subatomic particle: protons, neutrons and electrons. The protons and neutrons are concentrated at the centre of the atom, in the nucleus, while the electrons are well outside the nucleus (see 1.1). The table shows the mass and electric charge of each particle, relative to those of the proton.

Particle	Relative mass	Relative charge	Symbol
proton	1	+1	p
neutron	1	0	n
electron	$\frac{1}{2000}$	−1	e

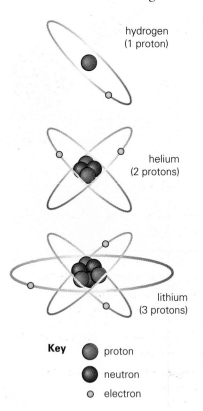

Key: proton, neutron, electron

An early theory of atomic structure was the 'plum pudding' model. This envisaged the atom as a positively charged sphere, throughout which the negative electrons were distributed like currants in a pudding! The discovery of radioactivity threw doubt on this model. The scientist Ernest Rutherford, while investigating the properties of alpha particles (one type of nuclear radiation), found that when they passed through thin metal foil a few were scattered at large angles from the direction of the incident beam. In 1911 he proposed that these results could only be explained if all the positive charge and most of the mass of an atom was concentrated in a very small nucleus. This was the basis for our modern model of the atom.

The number of protons the nucleus contains determines what the element is. Hydrogen atoms have one proton, helium atoms have two protons, lithium atoms have three protons and so on. Uranium atoms have 92 protons.

Although all atoms of a particular element have the same number of protons, they don't all have the same number of neutrons. The number of protons which an atom contains is called the **proton** (or **atomic**) **number**. The number of protons and neutrons together, is called the **nucleon** (or **mass**) **number**. Therefore:

number of neutrons = nucleon number − proton number

Isotopes

Atoms which have the same proton number (and so are the same element) but a different nucleon number (i.e. a different number of neutrons) are called **isotopes**. Isotopes which are unstable, and so radioactive, are sometimes called **radioisotopes**. Carbon has a number of naturally occurring isotopes. Atoms of carbon-12 are stable, but atoms of carbon-14 are radioactive.

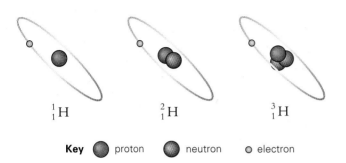

The three isotopes of hydrogen.

The number attached to the name of the element is the nucleon number of the isotope.

The use of symbols is often helpful:

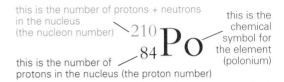

What types of nuclear radiation are there?

Nuclear radiation was found to consist of three types:

- **alpha** rays (or particles), which are helium nuclei
- **beta** rays (or particles), which are high-energy electrons
- **gamma** rays, which are electromagnetic waves with a shorter wavelength (higher frequency) than X-rays.

The table on the right summarises the nature of these rays.

Radiation	Relative mass	Relative charge	Symbol
alpha (α) particle	4	+2	$^{4}_{2}\text{He}$
beta (β) particle	$\frac{1}{2000}$	−1	$^{0}_{-1}\text{e}$
gamma (γ) ray	–	–	γ

What happens in the nucleus?

Polonium-210 breaks down spontaneously (decays) to form lead-206, emitting an alpha particle.

This reaction can be shown by a nuclear equation:

$$^{210}_{84}\text{Po} \longrightarrow {}^{206}_{82}\text{Pb} + {}^{4}_{2}\text{He}$$

Nuclear equations have to balance, just like chemical equations. This means that all the protons and neutrons have to be accounted for. Check this for the equations on this page.

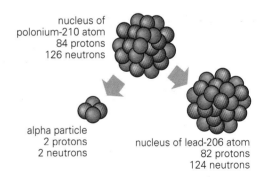

Alpha emission causes two protons and two neutrons to be lost from the nucleus – the proton number decreases by 2 and the nucleon number decreases by 4.

When carbon-14 decays, a beta particle is emitted from the nucleus and the element changes to nitrogen:

$$^{14}_{6}\text{C} \longrightarrow {}^{14}_{7}\text{N} + {}^{0}_{-1}\text{e}$$

When a beta particle is emitted, the nucleon number stays the same, but the proton number increases by 1. This is the result of a neutron being transformed into a proton and a high-energy electron.

When gamma radiation is emitted, there is no change in the number of protons or neutrons. Gamma emission often occurs after alpha or beta emission, due to a rearrangement of the nucleus to a more stable configuration.

QUESTIONS

1 Draw a diagram of an atom of $^{12}_{6}\text{C}$ and of $^{14}_{6}\text{C}$.

2 In the following list of radioisotopes, the chemical symbols have been replaced by the letters A–E.

$^{224}_{88}\text{A}$ $^{228}_{88}\text{B}$ $^{228}_{89}\text{C}$ $^{228}_{90}\text{D}$ $^{232}_{90}\text{E}$

a Which radioisotope possesses the most neutrons?

b Which are isotopes of the same element?

10.2 Properties and detection of nuclear radiation

Ionising radiation

When the radiation from radioactive sources collides with neutral atoms or molecules these can become charged or **ionised** – an effect first discovered by Marie Curie. The beam of radiation loses energy by this process – it is **absorbed** by the material through which it passes. X-rays are another form of ionising radiation.

When ionising radiation is absorbed by the molecules of living cells, cancer may result.

One of the ways that alpha, beta and gamma radiations can be distinguished is by their differing ability to penetrate materials.

- Alpha particles are highly ionising and are absorbed by a few centimetres of air. They can pass through a thin sheet of paper, but are absorbed by anything thicker, or denser.
- Beta particles are less ionising but more penetrating. They can pass through card and thin sheets of metal, but are absorbed completely by sheets of metal a few millimetres thick.
- Gamma rays have the properties of high-frequency electromagnetic waves. They are more penetrating than alpha or beta particles. Even a piece of dense material like lead several centimetres thick is unable to stop them completely. The thicker the material, the more radiation is absorbed. In addition, for a given thickness of material, the greater its density, the greater the absorption.

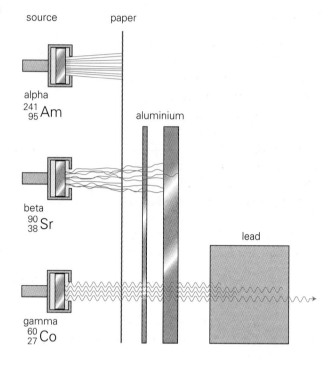

Absorption of alpha, beta and gamma radiation.

Storing radioactive materials

The radioisotopes used in schools are low-level sources (see 10.6). Even so, they are always stored locked away in lead-lined containers. Because the lead absorbs the majority of the gamma radiation along with all the alpha and beta radiation emitted, the danger from these ionising radiations is minimised.

When not in use, radioactive sources are stored in lead containers.

Detecting nuclear radiation

Most detectors work by indicating the arrival of energy, which may produce:

- exposure of photographic film
- ionisation, as in a Geiger-Müller tube and other devices
- increased current in a semiconductor device.

Photographic film is used to monitor the amount of radiation received by people working with radioactive materials or X-rays. A specially manufactured 'film badge' is worn during working hours. At regular intervals, the badge is developed and replaced. Once developed, because of the construction of the badge, both the amount and type of radiation received are revealed.

While photographic materials make an extremely useful kind of detector, they don't give an instantaneous picture of the amount or type of radiation. If this is needed, other types of detector are required. A **Geiger–Müller** (**GM**) **tube** is one such detector.

Ionising radiation is able to enter the GM tube through a thin mica window. The argon gas in the tube becomes ionised and a pulse of current passes between two electrodes. These pulses are detected and counted by connecting the GM tube to a suitable electronic counter. The information is sometimes displayed as the number of counts per second – the **count rate**. The greater the count rate, the greater the rate of ionisation and hence the greater the amount of radioactivity.

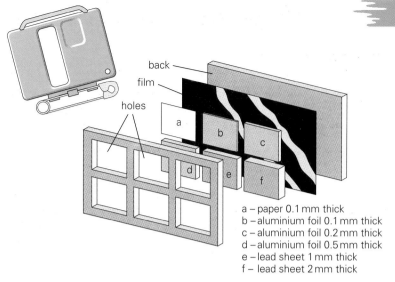

a – paper 0.1 mm thick
b – aluminium foil 0.1 mm thick
c – aluminium foil 0.2 mm thick
d – aluminium foil 0.5 mm thick
e – lead sheet 1 mm thick
f – lead sheet 2 mm thick

Film – no exposure to radiation

Film shows exposure to radiation

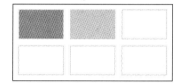

The construction of a film badge.

The whole counting device (the power supply to the tube, the tube itself and the pulse counter) is often referred to as a **Geiger counter**.

The type of radiation can be found, or its penetrating power investigated, by placing specially prepared sheets of card or metal in front of the mica window.

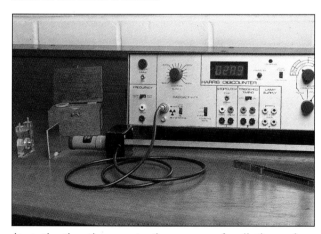

Investigating the penetrating power of radiation using a GM tube and counter.

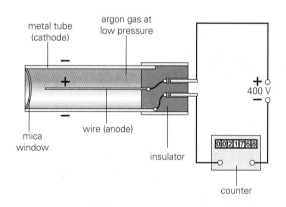

The construction of a GM tube.

QUESTIONS

1. Explain how the film badge illustrated is able to distinguish between different types of radiation.

2. Describe briefly how a Geiger counter could be used to identify the types of radiation being emitted by a radioactive source.

3. Why should tongs be used when handling low-level sources?

4. Why is an alpha source that has been ingested into the body much more dangerous than one outside the body?

10.3 Radioactive decay

Why does it happen?

The protons and neutrons in a nucleus are held together by a balance of forces. The electrostatic repulsion between the positively charged protons is balanced in a stable nucleus by the strong interparticle ('nuclear') forces that exist at the subatomic level. The most commonly occurring isotope of an element is the one that has the most stable nucleus. Radioisotopes have either too many or too few neutrons for stability. Broadly speaking, if there are too many neutrons, the interparticle attraction within the nucleus outweighs the electrostatic repulsion. If there are too few neutrons, the electrostatic repulsion of the protons takes over. These isotopes will undergo radioactive decay to a more stable nuclear arrangement.

Elements which contain more protons than lead – those with a proton number greater than 82 – are *only* able to form nuclei which are unstable. This means that *all* their isotopes are radioactive. Elements with a proton number greater than 92 are so unstable that they no longer exist naturally, although more than a dozen 'new' elements have been created artificially.

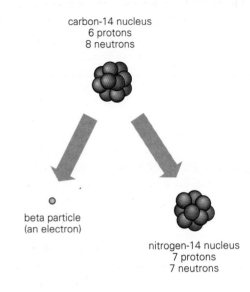

Carbon-14 is unstable because of its excess neutrons.

How fast is the decay?

Individual nuclei of a radioisotope break down, or decay, *spontaneously* – it is impossible to tell when a particular atom will decay. In a sample of the radioactive material, decay occurs *randomly* – for example, two pulses may be registered by a Geiger counter one after the other, and then a long time may elapse before the next pulse.

The average number of nuclei which break down each second, or the rate of decay, depends only on:

- what the isotope is – different radioisotopes decay at different rates
- the number of undecayed nuclei left in the sample.

The overall **rate of decay** is measured by the count rate. The rate is not altered by changing the physical conditions such as temperature or pressure, nor by different chemical conditions such as bonding.

As time passes, fewer and fewer of the original atoms in a sample will remain, so the rate of decay – sometimes called the **activity** – will decrease with time. The activity of a radioactive source is measured in **becquerels** (Bq), after the discoverer of radioactivity. A source which emits an average of 10 000 particles per second will have an activity of 10 000 Bq. Sources used for medical purposes may have an activity 10 thousand million times greater than the low-level sources in schools.

Half-life

For any given isotope, the time taken for a particular fraction of the nuclei present to decay is always the same. The **half-life** is the time taken for half of the nuclei present to decay.

The half-life of radium-221 is 30 seconds; so, if a sample had 1 billion of these nuclei at the start, then after 30 seconds, there would be just $\frac{1}{2}$ a billion of them left. After a further 30 seconds, the number would have halved again to $\frac{1}{4}$ of a billion, and after a further 30 seconds, would have halved again to $\frac{1}{8}$ of a billion. So after a period of time equal to three half-lives (1.5 minutes in the case of radium-221), $\frac{7}{8}$ of the nuclei originally present will have decayed.

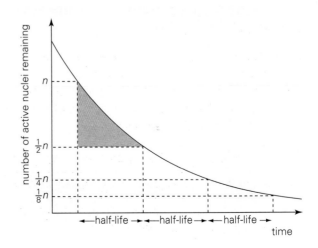

The half-life of different radioisotopes varies enormously – some have half-lives of billions of years, whereas others have half-lives measured in tiny fractions of a second! The table shows the half-lives of some naturally occurring radioisotopes.

Element and symbol	Proton number	Nucleon number	Half-life
uranium U	92	238	4.5 thousand million years
uranium U	92	234	250 000 years
polonium Po	84	210	138 days
radon Rn	86	222	3.8 days
polonium Po	84	214	0.00016 seconds

A useful analogy to help you understand the concept of half-life is as follows. If 200 coins are tipped onto a table, half the coins (100) would be expected to fall as 'tails', and half to fall as 'heads'. If the coins which had fallen as 'heads' are collected up and tipped onto the table again, half of them (50) would be expected to fall as 'heads'. After three such throws (equivalent to three half-lives), $\frac{7}{8}$ of the coins (175) will have been removed from the game.

In the analogy, as with real isotopes, it is impossible to say which specific coins will fall as a 'tail', or which nuclei will decay. Also, it would be most unlikely for *exactly* half of the coins to fall as 'tails' each time, or exactly half of the nuclei to decay in each half-life. However, statistically, this is what happens on average.

Measuring half-lives

In theory, the amount of radiation being emitted can be used as a measure of the amount of the isotope that remains, but in practice things are usually more complicated. This is partly because the 'daughter products' are frequently radioactive, and so these will therefore be emitting radiation as well.

Radon-220 is one substance whose half-life is straightforward to measure. This is because all of the daughter products have significantly different half-lives, and therefore have no appreciable affect on the rate at which the radiation level falls.

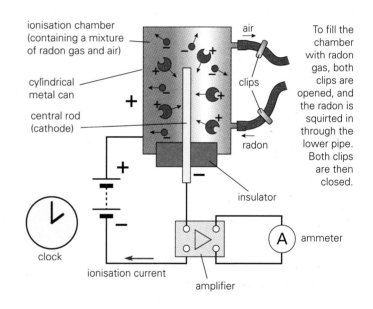

Measuring the half-life of radon gas with an ionisation chamber.

QUESTIONS

1 The following measurements were taken using an ionisation chamber and a sample of the radioactive gas radon-220. The current produced by the ionisation chamber is a measure of the amount of radon-220 remaining.

Time (seconds)	0	20	40	60	80	100
Current (mA)	0.95	0.72	0.55	0.42	0.32	0.28

Plot a graph of current against time, and calculate from it the half-life of radon-220.

2 Explain why elements with a proton number greater that 92 do not occur naturally.

3 If every time you tossed a pile of coins, you removed all those that landed as a 'head', after how many tosses would you expect to have just 10 coins left, if you started with 2560?

4 Uranium-234 has a half-life of 250 000 years. Sketch its radioactive decay curve. What fraction of a sample originally present would remain after a period of 5 million years?

10.4 Radioactivity and dating

The reduction in activity of a radioactive sample with time, whatever the outside conditions, is the principle behind the determination of the age of samples, or 'dating'.

Carbon dating

Plant and animal remains contain the radioisotope carbon-14. Its half-life of 5730 years makes it ideal for dating archaeological specimens.

The source of the carbon-14 is the atmosphere. Cosmic radiation causes the ejection of neutrons from nuclei in the atmosphere. Some of these combine with nitrogen-14 to form carbon-14. The proportion of carbon-14 in the atmosphere is thought to have remained constant at about 1 in every 10^{12} atoms of carbon. This is because the rate of neutron bombardment is thought to have stayed the same, and because the rate at which carbon-14 is being produced is exactly balanced by the rate at which it is decaying. A fixed proportion of the carbon incorporated into plants by the process of photosynthesis is therefore carbon-14, the rest being mainly carbon-12. The carbon-14 is passed on down the food chain.

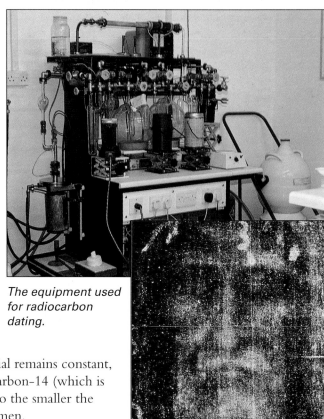

The equipment used for radiocarbon dating.

The ratio of carbon-12 to carbon-14 in living material remains constant, but when a plant or animal dies, the proportion of carbon-14 (which is no longer being replaced) falls as decay takes place. So the smaller the proportion of carbon-14 present, the older the specimen.

The use of carbon dating has featured in the news in recent years in several cases which have captured the public interest – for example, the investigation of the Turin Shroud, the 'Bogman', and the restoration of the flagship the *Mary Rose*.

A computer-enhanced image of part of the Turin shroud.

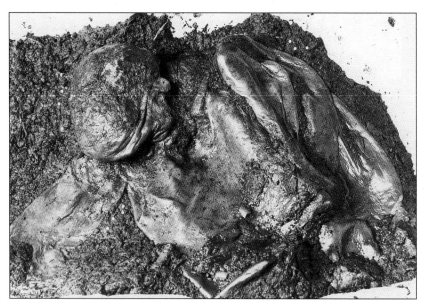

The 'Bogman', Pete Marsh.

Tradition had it that a shroud housed in the cathedral in the Italian city of Turin, had been used to wrap the body of Jesus. The shroud is marked with an impression of a body, rather like a photographic negative. No-one could explain how it got there, but it was supposed to be an impression of Jesus himself. After many years of discussion, the authorities finally decided to allow a small sample of the shroud to be carbon-dated. The results showed that the shroud couldn't have wrapped the body of Jesus, as it hadn't been made until long after he had died.

Decay chains and dating the Earth

When a large unstable isotope such as uranium-238 decays, new 'daughter' atoms are formed which themselves are unstable and after a time break down. In this way, by a series of steps, uranium-238 is gradually changed into the stable isotope lead-206. The first stage, the decay of uranium-238 into thorium-234, has by far the longest half-life (4.5 thousand million years). It is this step which controls the overall rate at which lead-206 is formed. If all the lead-206 now on Earth started out as uranium-238 then the ratio of the two isotopes today should tell us the age of the Earth. As their relative abundance is roughly equal, a rough figure of 4.5 thousand million years (one half-life) emerges as the age of the Earth. This figure is supported from other sources.

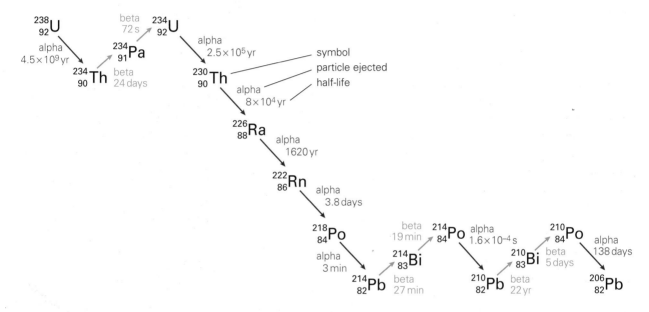

The uranium-238 decay series.

Dating igneous rocks

Up to about 0.01% of the total potassium present in igneous rocks can consist of the radioisotope potassium-40 (half-life 1.3×10^9 years). As the potassium-40 decays, there is a simultaneous build-up of the stable isotope argon-40. Provided the gaseous argon has remained trapped by the crystal structure, the ratio of potassium-40 to argon-40 can be used to determine the age of the rock.

QUESTIONS

1 The half-life of potassium-40 is over 1 thousand million years. It is found in plant and animal remains, but is not used to date archaeological specimens. Why not?

2 Carbon-14 is formed in the atmosphere by the bombardment of nitrogen by neutrons. Copy and complete the nuclear equation for this reaction:

$$^{14}_{7}\text{N} + ^{1}_{0}\text{n} \longrightarrow ^{14}_{6}\text{C} + ^{?}_{?}?$$

3 If the Turin Shroud had been shown to have been made before Jesus died, would the scientists have proved that it had been used to wrap the body of Jesus?

4 Suggest why carbon-14 is not used to date samples which are more than about 50 thousand years old.

10.5 Radioactivity and you

Radiation frequently has something of a bad press. You cannot see, smell or taste it, and it is linked to unpleasant things such as nuclear weapons and cancer. What are the facts?

The biological effects of ionising radiation

Ionising radiation can damage living cells. Alpha particles cause more ionisation than beta particles or gamma rays. Alpha particles are, however, normally unable to penetrate the non-living outer layers of skin, whereas beta particles and gamma rays penetrate further (see 10.2). So when alpha particles are outside the body, they are the least dangerous of the three types of radiation. If alpha particles are taken inside the body (such as into the lungs) where the cells are living, the position is reversed – they are highly damaging.

	Early or short-term	Later or long-term
High dose – as in accidents		
To whole body	Death in a matter of days or weeks	
To limited area of skin	Reddening of the skin as with sunburn	Possibility of skin cancer and other damage
Low doses – as from natural radiation – or high dose spread over a long time		
	No observable effects	Possible effects years later – cancers and hereditary diseases

Effects of ionising radiation on health. Source: NRPB.

Doses of radiation

There are two kinds of effects on health – those which show up more or less straight away and those which show up much later on in life, as the table shows.

To cause harm, radiation has to be absorbed. The absorbed dose is measured in terms of the energy absorbed per kilogram of body tissue. A more useful guide to the likely effect of the absorbed radiation is the **equivalent dose**, which takes into account the nature of the radiation. However, some organs are more sensitive to radiation than others, and this needs to be taken into account as well. The **effective dose** or simply **dose** represents the overall risk. Doses are measured in sieverts (Sv). Sieverts are quite large units, so microsieverts (μSv) are often used instead.

Background radiation

The one thing you can be sure of with ionising radiation is that you cannot avoid being exposed to it. We are continuously exposed to natural radiation from the Sun, the rocks that make up the Earth's surface, the air we breathe and the food we eat. Additionally, we are exposed to radiation as a result of human activities. Radiation from these sources is known as **background radiation** and always contributes to the reading shown on a Geiger counter.

Cosmic rays from the Sun and other stars are partially absorbed by the Earth's atmosphere. The higher you go, the less protection the atmosphere can give. A short-distance flight of about two hour's duration will expose you to an extra dose of about 10 μSv.

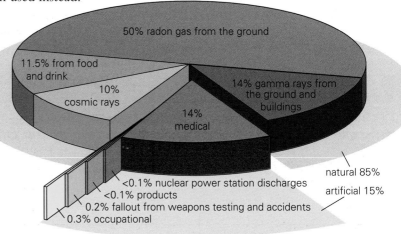

On average, each UK citizen receives a radiation dose of 2600 μSv a year. Source: NRPB.

The element uranium is widely distributed in the Earth's crust, and in building materials such as stone, bricks and concrete. The uranium breaks down by a series of steps into a stable isotope of lead (see 10.4). Some of the gamma rays from these reactions escape. Radon, a radioactive gas, is one of the daughter products. Some seeps out into the atmosphere, where it is dispersed.

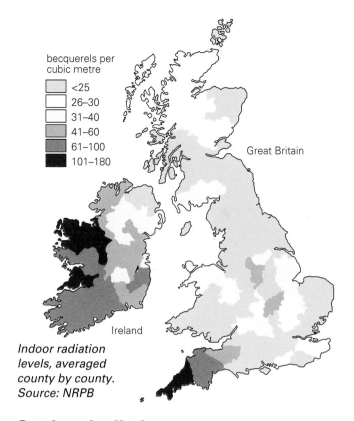

Indoor radiation levels, averaged county by county. Source: NRPB

In parts of the country the radiation levels are higher – for example, in Devon and Cornwall, where the underlying rock is granite which contains uranium. People who live in these areas are likely to be exposed to higher concentrations of radon if it seeps from the ground into their houses. How much gas builds up inside a particular building depends on a variety of factors. Forced ventilation of the space beneath the ground floor is sometimes recommended.

Moving around the country is unlikely to have any effect on the dose which you receive as part of your diet. On average, about 360 million radioactive potassium-40 atoms and 168 000 uranium atoms distintegrate inside each one of us every day of our lives.

In comparison to the average dose received from these natural sources, the average dose received from radioactive discharges from nuclear power stations and fallout from events such as the Chernobyl explosion is tiny.

Setting the limits

It is generally assumed that, however low the dose of radiation, there is some risk of harmful effects and that the risk is proportional to the dose. Guidelines therefore exist to minimise the unnecessary exposure to which a person is subjected. In addition, dose limits are set, being lower for the general public than for people who work with ionising radiation. For most of these workers, the radiation dose received each year from natural sources is greater than that received as a result of their work.

QUESTIONS

1. List the factors which would affect the levels of radon gas that could build up in your living room.

2. When talking about doses of radiation, why don't scientists simply use becquerels?

3. Estimate the additional annual dose a courier would receive if, during the year, she made 150 flights across Europe whose average duration was three hours.

4. 1g of potassium-40 contains approximately 1.5×10^{22} atoms.
   a. What would be the mass of 3.6×10^8 atoms (the average number to decay inside each person each day)?
   b. What would be the approximate mass of potassium-40 to decay in an average person over a period of fifty years?
   c. What would be the approximate mass of potassium-40 to decay inside the entire UK population (55 million people) in the course of one year?

5a. What is the average yearly dose that each person in the UK receives from natural sources? (Refer to the pie-chart opposite.)
  b. An average X-ray delivers a dose of 20 μSv. How many X-rays a year would be needed to equal the average yearly dose of radiation?

10.6 Radiation at work

Most people tend to think of radioactivity in connection with nuclear power or atomic bombs, but radioactive materials are used for a wide variety of purposes, in industry, in medicine, and even in the home! The isotope chosen for a particular job depends on a variety of factors, including its cost, its half-life, and the type of radiation emitted. In some of the examples given here, sealed sources of radiation are used. The radioactive material is sealed inside metal foil and, provided normal safety procedures are followed, there is no prospect of any escaping and 'contaminating' anything else. When open sources are used, for example in tracing techniques, isotopes with a short half-life are usually chosen.

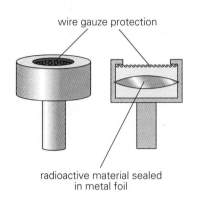

Radioactive material cannot escape from a sealed source in a school under normal usage.

Smoke alarms are sensitive, cheap and reliable.

Smoke alarms

You probably have a smoke alarm fitted in your home. Many detectors contain americium-241 which has a half-life of 458 years. The sealed source sends a stream of alpha particles to a sensor across a small air gap. Alpha particles are easily blocked, however, so if smoke fills the gap fewer of them reach the sensor. This change is used to trigger an electronic alarm circuit. A source which emitted beta particles or gamma rays would be unsuitable because they are too penetrating – the smoke would not stop them from reaching the sensor.

Thickness gauges

Manufacturers of sheet materials such as paper or plastic need to know that the continuous sheets produced by their machines are of uniform thickness. Beta particles are penetrating enough for a significant quantity to pass through such materials (alpha particles would be stopped completely). If a sealed beta source is placed opposite a sensor that measures the proportion of particles getting through, a constant value will show a constant thickness. If the sheet is too thin, the value will rise because of a decrease in absorption; if it is too thick, the value will fall. The information is monitored continuously and warnings are triggered if the thickness varies by more than a predetermined amount.

Manufacturing plastic sheeting.

Following the flow

It is impossible to tell how water or other liquid is flowing through a closed piping system under normal conditions. If a gamma source such as iridium-183 (half-life 54 minutes) is fed into the water of a cooling system, its movement can be monitored with a Geiger counter held alongside the pipework and any leaks traced. Alpha or beta particles would be absorbed by the metal of the pipework.

Similar techniques have been used to follow the flow of underground watercourses and also to check how fast water is taken up by plants.

Medical tracers

Modern radiation detectors are so sensitive that they can measure very low levels of radiation that present no perceived danger to us. Because of this, a very small amount of an appropriate radioactive substance can be attached to or 'tagged' to carrier chemicals and safely injected into the bloodstream, and then its radiation monitored as the chemicals pass through the body. One such radioactive 'tracer' is technetium-99, a gamma emitter with a half-life of 6 hours.

This technique is used to check how a particular part of the body is functioning. The liver, for example, gets rid of unwanted chemicals from the body. If an appropriate chemical is tagged, the liver should emit radiation which can be detected. If a part of the liver does not, then that part is not functioning correctly and may need medical treatment.

Tracers used have a short 'biological half-life'. This is the time it takes for half of it to pass through or be excreted by the body.

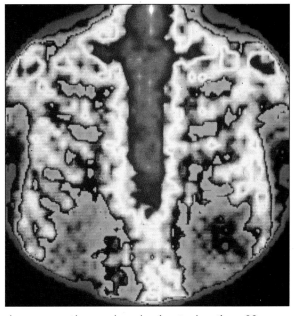

Images can be produced using technetium-99.

Treating cancer

It perhaps seems odd, but gamma rays – a possible cause of cancer – can also be used to treat it. Cancer cells are more easily damaged by ionising radiation than healthy cells, and high, localised doses can kill them without severe damage to normal cells. In one form of **radiotherapy**, gamma rays from a cobalt-60 source are focused onto the cancerous tumour, each dose being precisely calculated beforehand. It is safer but not so effective to use high-energy X-rays as, unlike a source of gamma rays, an X-ray machine can be switched off when not required.

QUESTIONS

1. Why are open sources potentially far more dangerous than sealed sources?

2. Explain why the radioactive substances used as tracers in medical tests:
   a. have a short half-life
   b. have a short 'biological half-life'
   c. are beta or gamma emitters, and
   d. are given in small quantities.

3. a. Why is an alpha-emitting source used in a smoke alarm?
   b. Why is there no danger from the alpha source in a smoke alarm?
   c. How long will it take for the activity of an americium-241 source to fall by half?
   d. Why should you hoover your smoke alarm every so often?

10.7 The source of nuclear energy

Mass and energy

In the simple model of the atom described in 10.1, each proton and neutron has a relative mass of 1. When accurate measurements are made, however, the surprising result is that atoms have a mass that is less than the sum of their parts! In 1905, while working on his theories of relativity, Albert Einstein suggested that mass was equivalent to energy and that the two were linked by the formula:

$$E = mc^2$$

where c is the velocity of light.

The reduced mass of nucleons when bound in a nucleus is accounted for by the binding energy. Because of the interparticle attraction, the nucleons have lower (potential) energy when bound together and hence a lower mass.

The key to understanding nuclear energy is that *any reaction that gives an overall mass loss will liberate energy*. The decay of uranium-235 generates about 500 million times as much energy per atom as any of its chemical reactions! For each gram of mass lost in a nuclear reaction, 9×10^{13} joules of energy are produced.

Fission

Uranium-235 decays slowly through a series of stages, ending up as lead. This steadily liberates energy over thousands of millions of years. The production of energy is speeded up if neutrons bombard the nucleus. When this happens, instead of forming thorium-231 and an alpha particle, the uranium-235 nucleus is usually split in two. This is an example of **nuclear fission**. There are a variety of ways in which the split of uranium-235 can occur, so a variety of daughter atoms will be produced, which themselves are generally radioactive. In practice, certain splits occur with a much greater frequency than others. Amongst these is one in which strontium-90 and xenon-143 are formed:

$$^{235}_{92}\text{U} + ^{1}_{0}\text{n} \longrightarrow ^{90}_{38}\text{Sr} + ^{143}_{54}\text{Xe} + 3\,^{1}_{0}\text{n} \text{ plus energy}$$

The half-lives are: uranium-235, 710 million years; strontium-90, 28.1 years; xenon-143, 1.0 seconds.

The world's first atomic bomb had an explosive power equivalent to 10 000 tonnes of TNT.

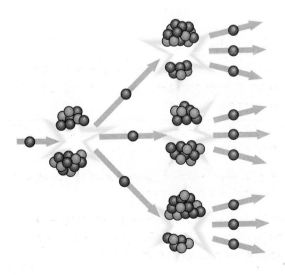

A chain reaction.

As each uranium atom is split, further 'fission neutrons' are liberated, which themselves may be capable of splitting further uranium atoms, thereby starting a **chain reaction**. Under certain circumstances, a virtually instantaneous, massive outburst of energy can result – an atomic explosion.

In naturally occurring uranium compounds, less than 1% of the uranium is uranium-235. The uranium-235 atoms are therefore widely spread and few of the neutrons produced by fission reactions cause further fission. In small quantities of concentrated uranium-235, the loss of neutrons through the surface is sufficiently high to cause no problems. But once a certain critical size is reached, sufficient internal collisions will occur to trigger a chain reaction and consequent explosion. So to make a simple atomic bomb, all that is needed is a critical mass of 'enriched' uranium-235 (about the size of a cricket ball) and the technology to make it work!

Controlling fission energy

All the world's nuclear power stations produce energy by nuclear fission, mostly of uranium-235. The nuclear reactions have to be controlled, so that an explosive release of energy doesn't occur. This is achieved by using boron steel rods, which are able to absorb neutrons. The rate at which fission occurs is controlled by raising or lowering these control rods in the core of the reactor. Uranium-235 splits when bombarded by *slow* neutrons, so the energetic fission neutrons are slowed down by a 'moderator'. In some types of reactor the moderator is graphite – the neutrons lose energy when they collide with the carbon atoms.

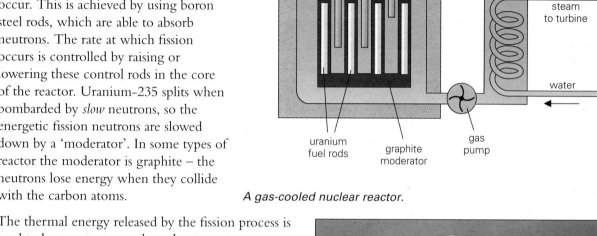

A gas-cooled nuclear reactor.

The thermal energy released by the fission process is used to heat water to produce the steam to turn a turbine (see 9.2). There have been, and continue to be, many arguments about the safety of nuclear power stations, both in their operation and in the methods of disposal of the radioactive waste produced. 'High level' waste, resulting from reprocessed spent nuclear fuel, may remain active for up to a million years, so its containment poses the greatest technological problem.

The new reactor at Sizewell in Suffolk.

Low-level nuclear waste from Sellafield is currently stored above ground in specially designed containers.

QUESTIONS

1. Where does the energy that is converted into electrical energy in a nuclear power station come from?

2. Explain why chain reactions don't occur in uranium ores.

3. Why do you think making an atomic bomb isn't as simple as it may seem?

4. If a nuclear power station was 40% efficient and had an output of 500 MW, what would be the daily mass loss of the nuclear fuel?

SECTION E: QUESTIONS

1a A teacher used a Geiger counter to investigate the radiation given out by a radioactive source. Before she did anything else, she started the counter and then stopped it after one minute. A count of 18 was shown.
Why did the counter show a reading, even though no radioactive source was being used?

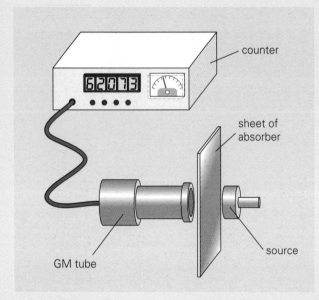

b The teacher then placed the radioactive source a few millimetres in front of the Geiger-Müller tube and took readings off the counter at one-minute intervals. She then placed a variety of absorbers in turn between the source and the tube, taking further readings as she did so. Her results are shown in the table.

Absorber used	Average counts per minute
none	1973
card 1 mm thick	1216
lead 1 mm thick	21

State whether or not each of the following radiations – *alpha*, *beta* and *gamma* – is given out by the source, giving reasons for your answer in each case.

2 By referring to the proton numbers and nucleon numbers, calculate the number of alpha and beta particles emitted when an atom of $^{238}_{92}U$ decays (by a series of steps) to the stable isotope of lead, $^{206}_{82}Pb$. Explain your calculation, and check your answer by referring to 10.4.

3 The half-life of radium-226 is 1620 years.
a What fraction of a given sample will remain after 4860 years?
b What fraction will have decayed after 8100 years?

4 The three stable isotopes of oxygen (proton number 8) have nucleon numbers of 16, 17 and 18. Draw diagrams of each of these three isotopes.

5 As each atom of $^{214}_{82}Pb$ decays, a beta particle is emitted, and an atom of bismuth is formed.
a What does the number 214 represent?
b What does the number 82 represent?
c The isotope of bismuth can be written as $^{x}_{y}Bi$. What are the values of x and y?
d Write an equation to show what is happening.

6a What is the biggest natural source of ionising radiation?
b Which of the natural sources of radiation is likely to vary the least up and down the country? Justify your answer.
c List three ways in which ionising radiation could be received during a course of medical treatment.

7 The isotope americium–241 is used in smoke detectors. It has a proton number of 95.
a How many protons and how many neutrons are there in each atom of americium–241?
b What type of radiation does americium–241 emit?
c Explain why there is no naturally occurring americium present in the Earth's crust.
d Americium–241 has a half-life of 458 years. Sketch a decay curve, and estimate how long it will take for the activity of the sample contained in a smoke detector to fall to:
  i 75% of its initial value,
  ii 25% of its initial value, and
  iii 10% of its initial value.

8a Describe the short-term and long-term effects of being exposed to:
  i low doses of radiation,
  ii high doses of radiation.
b Explain why people who work with radiation wear film badges.

9 Some watch manufacturers used to use radioactive paint to make the face and hands luminous. How could you tell, without using any special equipment, whether a luminous clock face had been painted with a radioactive paint or not?

10a Why are people in Devon and Cornwall more likely to be exposed to higher levels of radon gas than people living in Norfolk?

b Use the diagram to explain how the levels of radon gas within an existing building can be permanently reduced.

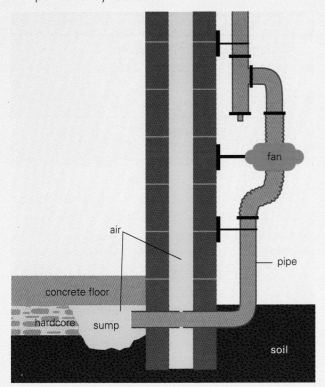

Reducing levels of radon gas in existing buildings.

c Many modern shopping centres have polished granite floors. Explain why there are unlikely to be problems arising from any radon gas which is emitted from the granite.

11a Explain what is meant by a chain reaction.

b What prevents a chain reaction from getting out of control in a nuclear reactor?

12a What is the origin of the carbon–14 found in living matter?

b Explain why the level of carbon–14 present in living matter remains constant, but starts to fall once death has occurred.

c Explain how carbon–14 can be used to determine the age of certain archaeological remains.

13a What are the differences between an alpha particle, a beta particle and a gamma ray?

b Copy and complete the equations to show how lead–210, and radon–222 decay.

$$^{210}_{82}\text{Pb} \rightarrow {}^{?}_{83}\text{Bi} + \ldots$$

$$^{222}_{86}\text{Rn} \rightarrow {}^{?}_{84}\text{Po} + \ldots$$

14 The chart shows the average annual radiation dose received by different groups of people in the UK.

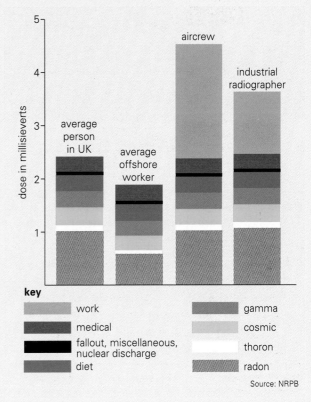

a List the sources of radiation that you cannot avoid being exposed to.

b List any other sources of radiation that you may be exposed to in the next 12 months.

c Suggest a reason why the average annual dose received from radon gas is lower for offshore workers.

d Suggest why the average annual dose received by aircrew while working is greater than that received by industrial radiographers.

CHAPTER 11: THE EARTH AND BEYOND

11.1 Our solar system

Our solar system consists of the Sun and the space surrounding it. Most of the space is virtually empty, but within it there are the nine known planets and their moons. In addition, there are many millions of smaller objects: the asteroids and comets. All of these bodies orbit the Sun.

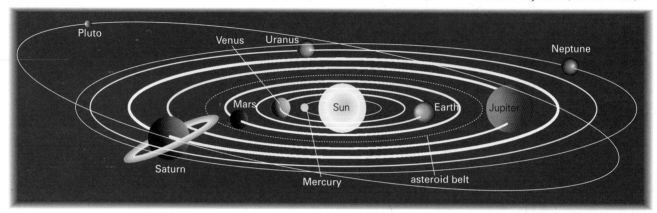

The solar system (not to scale).

The table shows some physical data for the planets.

Planet	Average distance from Sun (km)	Time to orbit the Sun (years)	Time to rotate through 360° on its axis*	Equatorial diameter (km)	Mass (kg)
Mercury	58×10^6	0.241	58.7 days	4.878×10^3	3.3×10^{23}
Venus	108×10^6	0.615	243 days	12.10×10^3	48.6×10^{23}
Earth	150×10^6	1.00	23.9345 hours	12.76×10^3	59.7×10^{23}
Mars	228×10^6	1.88	24.6 hours	6.790×10^3	6.4×10^{23}
Jupiter	778×10^6	11.9	9.8 hours	143×10^3	18990×10^{23}
Saturn	1427×10^6	29.5	10.2 hours	120×10^3	5684×10^{23}
Uranus	2870×10^6	84.0	17.9 hours	52×10^3	870×10^{23}
Neptune	4497×10^6	165	19.2 hours	49×10^3	1028×10^{23}
Pluto	5913×10^6	248	6.39 days	2.4×10^3	0.1×10^{23}

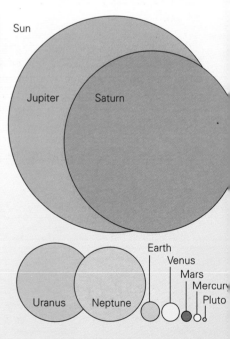

The relative sizes of the planets and the Sun. Planetary sizes and separations are rarely drawn to scale because of the huge differences in dimensions.

*The length of a planet's 'day' depends on both the direction and the rate at which it is rotating on its axis, and the rate at which it is orbiting the Sun. At 24 hours the average **solar day**, or 'day' on Earth, is approximately 4 minutes longer than the **sidereal day** – the time the Earth takes to rotate once on its axis.

Different parts of the gas giants rotate at different rates. The times of rotation quoted here are those at the equator.

Discovering the planets

Under the right conditions, all the known planets except Neptune and Pluto can be seen with the naked eye. However, Uranus was not identified until 1781 when William Herschel observed it through a telescope. Astronomers subsequently discovered that its path was not as they would have expected. They thought it was being affected by the gravitational pull of a yet undiscovered planet. In 1845, John Adams and Urbain le Verrier independently calculated where they thought the undiscovered planet would be found. The following year, the planet was observed and named Neptune. A ninth planet, Pluto, was discovered in 1930.

The planets all move in the same direction

It is generally believed that our solar system formed nearly 5000 million years ago, from a cloud of interstellar gas and dust. Gravity caused the cloud to contract, with over 99.9% of the mass forming a young star (the Sun), at the centre of a flattened spinning disc.

The planets formed within the spinning disc, and as a result, they all orbit the Sun in the same direction. The solar system is normally drawn showing the Earth with its north pole at the top. When it is viewed like this (from a position above our north pole) the planets move in an anticlockwise direction.

All of the planets except Pluto orbit in the same plane (to within a few degrees), as would be expected from the 'spinning disc' model. Pluto's orbit, however, is tilted by about 17° to the plane of the Earth's orbit. This has led to speculation about Pluto's origin.

The orbits of the planets are **elliptical** (elongated) rather than circular. This means that their distance from the Sun is continually changing. The average distance from the Sun to Pluto is much greater than the average distance to Neptune (see the table opposite), but because of the highly elliptical orbit of Pluto, there are times when Pluto is closer to the Sun than Neptune. The further the planet is from the Sun, the greater the time it takes to orbit it.

The distance from the Earth to the other planets depends on where they are in their orbits. Sometimes we are closest to Mercury, sometimes to Venus, and sometimes to Mars. When Earth and Mars are on opposite sides of their orbits, they are five times further apart than they are when they are on the same side.

QUESTIONS

Refer to the table of data opposite.

1. Calculate the number of times each planet will orbit the Sun, in the time that it takes for Pluto to orbit once.

2. How many times greater or smaller than the Earth's are the diameters and masses of the other planets? Make a table.

3. Calculate the approximate volume and average density of the Earth and Saturn. (Assume the planets are spherical; volume of a sphere = $\frac{4}{3}\pi r^3$; density = m/v.)

4. Explain, with diagrams, why Venus isn't always the closest planet to Earth.

11.2 The planets

Observing the sky at night

As far as any ordinary observer on Earth can see, the stars have a fixed position in space, with the only motion that is visible being their movement across the sky (in much the way that the Sun seems to move) as the Earth rotates on its axis. In order to make the recognition of individual stars more straightforward, astronomers have divided the sky into the regions we call the **constellations**. Each constellation consists of easily recognisable fixed patterns of stars, the shapes of which reminded the observers of particular animals or people. The 12 constellations of the 'zodiac' (those in a narrow belt in the plane of the Earth's orbit) were among the earliest to be named, by Babylonian astronomers before 2000 BC.

When daily observations of the sky are made, a small number of objects that look like stars are seen to slowly change position within certain of the constellations. The ancient Greeks named them *planets* (which means 'wanderers'). The changes in position are too small to notice as you are looking, but become apparent over a period of several days. The path of a planet through the constellations can be explained by considering the relative orbital motion of the planet and the Earth against the background of 'fixed' stars (see 11.4).

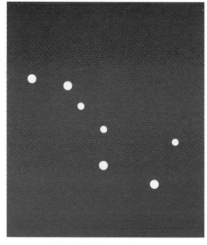

The Plough is one of the easiest groupings of stars to identify. Its position and orientation in the sky changes as the Earth spins on its axis.

Planets produce no light of their own

We can see the Sun because it is continually radiating the energy (some of which is light energy) which is produced at its core by nuclear fusion reactions (see 11.5). The planets have too small a mass to produce fusion energy in this way – Jupiter, the most massive, is at least five times too small. The reason the planets can be seen in the sky is that they reflect light falling on them from the Sun. The planets are much further away from us than our Moon, so although most of them are much larger, they actually look very much smaller. Mercury and Venus, which are between the Sun and the Earth, show phases rather like the Moon (see 11.3).

The energy from the Sun

The further a planet is from the Sun, the less radiant energy per square metre it receives. If the distance from a planet to the Sun were doubled, it would receive a quarter of the amount of energy. The surface of Venus is hotter than otherwise might be expected because of the greenhouse effect caused by its atmosphere of carbon dioxide and sulphuric acid.

This image of the surface of Venus was constructed with data gathered by the Magellan radar-mapping spacecraft.

'Rocky midgets' and 'gas giants'

The four innermost planets (Mercury, Venus, Earth and Mars) are sometimes called 'rocky midgets', because they are relatively small and are made almost entirely of rock. They are all thought to have a layered structure similar to that of the Earth (see 7.5), and all show evidence of past volcanic activity. The remainder of the planets (except Pluto) are called 'gas giants'. They are thought to be made up largely of a mixture of hydrogen and helium, with small rocky cores at their centres.

Less is known about Pluto than the other planets. It is smaller than our Moon and has a surface of frozen methane. A suggestion has been made that it might be an escaped moon of Neptune.

Jupiter and its moons Io and Europa, photographed by the Voyager 1 spacecraft.

The surface of Mars photographed by the Viking One spacecraft.

QUESTIONS

1. Why is the surface temperature of Venus higher than that of Mercury, even though it is further from the Sun?

2. What factors will affect the surface temperature of the 'rocky midgets'?

3. How can a planet be distinguished from a star without the aid of a telescope?

4. Most books quote very similar or identical figures for the diameters of the inner four planets (the rocky midgets). This is often not the case, though, when it comes to the gas giants. Suggest a reason why this should be.

11.3 Periodic changes

Why do we have seasons?

When the part of the Earth you are on is facing towards the Sun it is day-time, and when it is facing away from the Sun it is night-time. Because of the Earth's rotation, the Sun appears to move across the sky during the day, rising in the east and setting in the west. As well as rotating on its axis, the Earth also orbits the Sun. It takes one year for the Earth to go round the Sun once. The Earth's axis is not perpendicular to the plane of the orbit – it is tilted at an angle of 23.5°. As the Earth orbits the Sun, judged against the background of stars, the direction in which the axis points scarcely changes (it changes by about 1° every 70 years). As a result, the north pole points towards the Sun on one side of the orbit, and away from the Sun on the opposite side of the orbit. When the north pole points towards the Sun, it is summer in the northern hemisphere; when it points away from the Sun, it is winter.

When the north pole is pointing away from the Sun, the Sun is lower in the sky at mid-day in the northern hemisphere than it is when the north pole is pointing towards the Sun. The Sun's energy is spread over a larger area – so we get less of it, and it is colder. We also get less energy for another reason: there are fewer hours of daylight in the northern hemisphere when the north pole is pointing away from the Sun.

If the Earth's axis wasn't tilted, we wouldn't have the changing seasons or the changing hours of daylight.

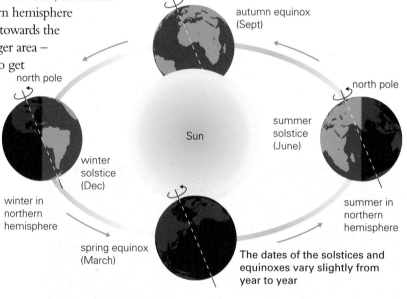

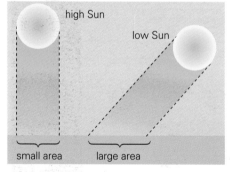

How much daylight do we get?

At the times of the spring and autumn **equinoxes**, night and day are of equal length. In the northern hemisphere, the amount of daylight increases each day between the day of the winter solstice (the 'shortest' day) and the day of the summer solstice (the 'longest' day), and then decreases again. The actual amount you get each day depends on where you live. In the UK, someone living north of you will have more hours of daylight in the six months between the spring and autumn equinoxes, but fewer hours for the remainder of the year. In the south of England, the 'longest day' has about twice as many hours of daylight as the 'shortest day'.

Path of the Sun at different times of the year.

The Earth's Moon

The diameter of our Moon is about a quarter of that of the Earth. Its mass is about 100 times smaller, and is insufficient for the Moon to have retained an atmosphere. It is in an elliptical orbit (in an anticlockwise direction) around the Earth. The plane of the Moon's orbit is tilted at an angle of about 5° to the plane of the Earth's orbit around the Sun. The distance from the centre of the Earth to the centre of the Moon varies between 356 400 km and 406 700 km, the average distance being 384 400 km.

The Moon takes 27.3 days to go around the Earth once. This is exactly the same amount of time that it takes for the Moon to rotate once on its axis. Because these rates are the same, the same face of the Moon always points towards the Earth. The only way to see the 'back' of the Moon is to observe it from space.

The first image of the back of the Moon was obtained by the Russian Luna 3 spacecraft in 1959.

The appearance of the Moon changes continuously over a $29\tfrac{1}{2}$-day cycle, resulting in the **phases** of the Moon. The reason that we can see the Moon at all, is because, like the planets, it reflects light from the Sun. The side of the Moon which is facing the Sun is always lit up. The amount of the lit face which can be seen from the Earth depends on the position of the Moon in its orbit. When the Moon is on the opposite side of the Earth to the Sun, all of it can normally be seen (a full Moon). When it is between the Earth and the Sun, none of it can be seen (a new Moon).

Full moon (as viewed from Earth).

Phases of the Moon.

QUESTIONS

1 Why is it summer in Australia when it is winter in the UK?

2 Why does the Sun never set at the north pole during June, and never rise during December?

3 Why are we unable to see the 'back' of the Moon?

4 Given that the Moon orbits the Earth in 27.3 days, why are there 29.5 days between each new Moon? (Hint: take into account the fact that the Earth and the Moon are also orbiting the Sun.)

11.4 What revolves around what?

For many thousands of years, people believed that the Earth was at the centre of the universe, and that everything else revolved around it. The daily rising and setting of the Sun, Moon and stars, seemed to indicate that this was so. However, there was a major problem in explaining the apparent motion of the planets across the 12 constellations of the zodiac, because, every now and again, the path of a planet would appear to 'backtrack' upon itself. Clearly, the planets weren't revolving simply around the Earth. One explanation was developed by the Egyptian astronomer Ptolemy, who lived during the second century AD. He successfully matched observed planetary motion with a model in which a planet revolved in a complex, 'looped' path around the Earth.

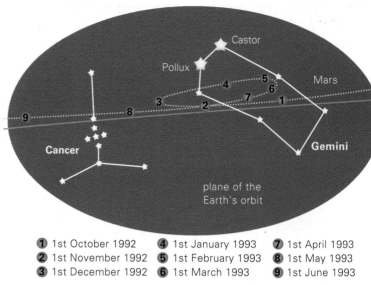

① 1st October 1992　④ 1st January 1993　⑦ 1st April 1993
② 1st November 1992　⑤ 1st February 1993　⑧ 1st May 1993
③ 1st December 1992　⑥ 1st March 1993　⑨ 1st June 1993

The path of Mars, through Gemini in 1992/93.

It was not until the early 1500s that this view was seriously challenged. A Pole named Nicolaus Copernicus developed a different explanation (the one that most people now believe), in which all the planets, including the Earth, are in motion around the Sun. His explanation was much simpler than that put forward by the ancient astronomers, but it was not popular with the church authorities.

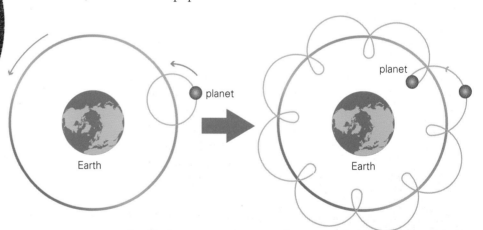

Nicolaus Copernicus, 1473–1543.

The ancient astronomers' explanation of the paths of the planets.

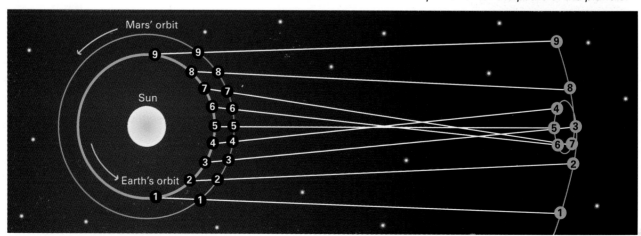

Copernicus's explanation of the apparent path of the planet Mars, when viewed against the background of stars.

Evidence for the Sun-centred model

When telescopes were invented in 1608 astronomers were able to start collecting the evidence that would finally convince people that Copernicus was right and Ptolemy was wrong. Galileo was the first to make astronomical observations using a telescope. His discovery of moons in orbit around Jupiter was the first evidence that objects existed which definitely did not move around the Earth.

In the eighteenth century, the British astronomer James Bradley produced the first evidence to support the idea (which by then was becoming more commonly held) that the Earth is moving around the Sun.

Jupiter and its four largest ('Galilean') moons are easy to spot with a small telescope.

The bright star Gamma Draconis passed almost directly overhead each night. Bradley observed the star regularly and noticed that, over the course of a year, its position appeared to change in a cyclical manner. He eventually realised that these tiny but measurable changes were due to the movement of the Earth as it orbited the Sun.

He reasoned that if the Earth was moving, then by the time light which had entered the top of his telescope reached the eyepiece, the Earth (and therefore the eyepiece) would have moved to one side. The direction in which the eyepiece moved would change throughout the year as the Earth orbited the Sun and, as a result, so would the apparent position of the star.

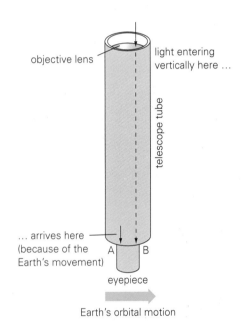

This apparent displacement of the position of a star due to the Earth's orbital motion is now known as 'stellar aberration'.

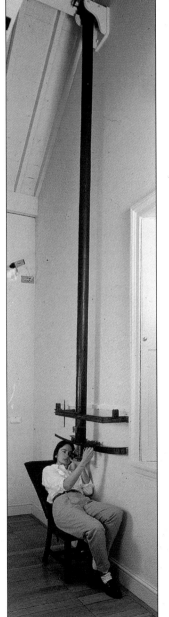

James Bradley discovered stellar aberration with this telescope.

QUESTIONS

1. How does the movement of the planets across the sky differ from that of the stars?

2. How can the observed movements of the planets be accounted for?

3a. Calculate how far the Earth will have moved around its orbit in the time that it takes for light to travel through a 4 m long telescope. (The speed of light is 3×10^8 m/s; the average speed of the Earth in its orbit is 29800 m/s.)
   b. How does this compare with the maximum distance the telescope could have moved as a result of the Earth's rotation? (Assume the Earth's diameter is 1.3×10^7 m.)

11.5 Stars and energy

Our Sun is one of thousands of millions of stars that make up the **galaxy** known as the Milky Way, which, in turn, is one of many thousands of millions of galaxies that make up the universe. Stars emit energy in the form of electromagnetic waves which travel out into space.

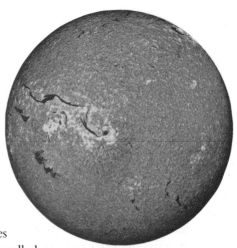

The Sun. **Never look directly at the Sun**, with or without a telescope, as blindness can result.

What is a light-year?

The universe is so large that it is often difficult for us to understand just how extensive it is. The distances between stars are hundreds of thousands times the distances between the planets of our solar system, and the distances between galaxies may be millions of times the distances between stars. Astronomical distances are so large, that a unit of distance called the **light-year** is used. One light-year is equal to the distance light (or other electromagnetic radiation) will travel through empty space in one year. Since light travels at a speed of 3×10^8 m/s, one light-year is equivalent to approximately 9.5×10^{15} m (9 500 000 000 000 000 m)! The nearest star to us (excluding the Sun) is Proxima Centuri, which is about 4.3 light-years away.

Where does the energy come from?

The Sun, like other stars, produces energy because, at its core where the temperature is extremely high (about 15 million degrees Celsius), **nuclear fusion** is taking place. The nuclei of hydrogen fuse to form the nuclei of helium and energy is released. During these 'thermonuclear reactions', which can only take place at such high temperatures, mass is converted into energy. Every kilogram of mass which is converted produces 9×10^{16} joules of energy: enough to run 250 million 100 W electric light bulbs for their lifetime of about 1000 hours each! Energy produced in this way is called fusion energy. A very small loss of mass produces a huge amount of energy compared to the energy released from a similar mass by a chemical reaction such as combustion.

Hydrogen can be converted into helium in a number of different ways. In the three-stage process shown here, the overall effect is that four hydrogen nuclei (each consisting of a proton) have combined to form a helium nucleus (consisting of two protons and two neutrons), plus some basic particles called positrons and neutrinos, and some energy. During the process, two protons are converted into neutrons, and there is an overall mass loss. The difference in mass between the starting and end products accounts for the energy released (see 10.7).

The conversion of hydrogen to helium.

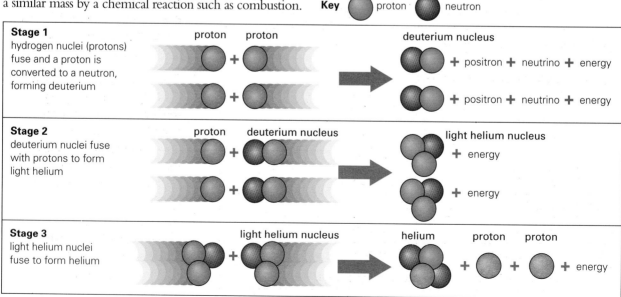

Why do the nuclei fuse?

At the very high temperatures and pressures in the core of the Sun, matter exists as **plasma**. In plasma, most nuclei exist on their own without their electrons. The nuclei have a lot of kinetic energy, and move around at very high speeds. They collide with one another frequently, overcoming the electrostatic repulsion between their positive charges, and react together – fuse – to produce a more stable nuclear state.

The protons and neutrons in a nucleus are held together by very strong nuclear forces, which come into play when the protons and neutrons are very close together. These attractive forces more than offset the repulsion of the positively charged protons.

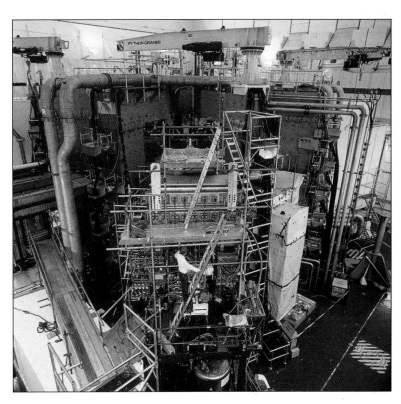

Scientists at work on Joint European Torus in Oxfordshire.

The Sun's gravity prevents the plasma from floating away. Another way of containing plasma is in a magnetic ring known as a torus. Scientists in Oxfordshire are using a torus to produce energy in a controlled way from nuclear fusion. At the moment, the amount of energy that can be safely produced is minute. Fusion energy is produced when a hydrogen bomb explodes.

Why we need the Sun

The energy produced by the Sun is essential for life on Earth. The Sun's energy keeps the Earth's surface at a temperature at which complex life forms are able to exist. Plants convert light energy from the Sun into chemical energy, and store it in the form of sugar and starch, to be ultimately used in food chains. The energy contained in fossil fuels came from the Sun millions of years ago.

Some facts about the Sun

Mass	2.2×10^{30} kg
Rate of loss of mass	4.2×10^9 kg/s
Life expectancy	10^{10} years
Time taken for energy to reach Earth	500 s (8.3 minutes)
Surface temperature	5700°C
Internal temperature	1.5×10^7 °C
Approximate composition	hydrogen (71%), helium (27%)
Distance to the centre of the galaxy	30 000 light-years

QUESTIONS

1. In what form does the Sun's energy reach the Earth? Does the Earth receive any of this energy at night?

2. The energy produced by a loss of mass in a nuclear reaction can be calculated from the equation $E = mc^2$, where energy E is in joules, mass m is in kilograms, and c is the velocity of light, 3×10^8 m/s.

   By referring to the table of data on the left, calculate how much energy the Sun produces each second.

11.6 The evolution of stars

From cloud to star

Stars and planets start life as a massive cloud of gas and dust, or **nebula**. Such a cloud would typically consist of about 75% hydrogen, 24% helium, and a smattering of other elements. The gravitational pull of the particles on one another causes them to move towards their common centre. Over time, the density and pressure of material at the centre increases. This denser region is called a **protostar**. As the particles 'fall' towards the centre, the temperature rises as their gravitational potential energy is converted into thermal energy (heat). When the temperature in the core of the protostar is high enough, thermonuclear reactions commence – hydrogen is converted into helium – and the star is born.

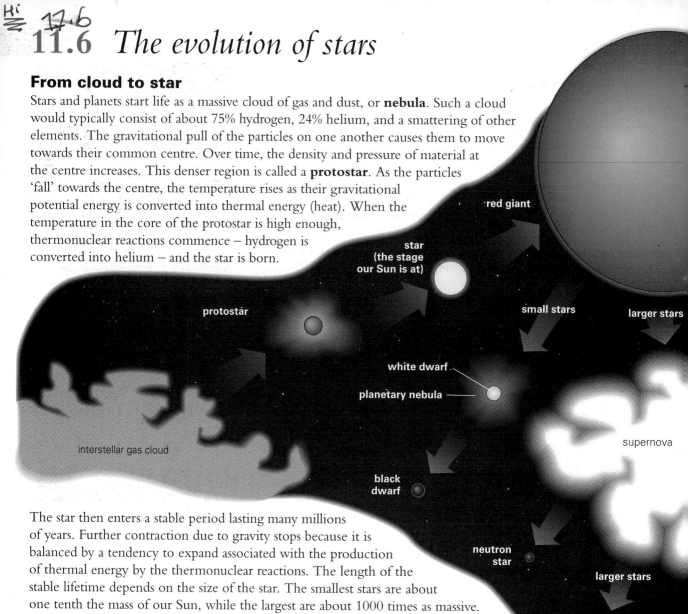

The star then enters a stable period lasting many millions of years. Further contraction due to gravity stops because it is balanced by a tendency to expand associated with the production of thermal energy by the thermonuclear reactions. The length of the stable lifetime depends on the size of the star. The smallest stars are about one tenth the mass of our Sun, while the largest are about 1000 times as massive. Our Sun is about halfway through its life expectancy of 10 000 million years.

The formation of red giants

Once all the hydrogen in the core has been converted into helium, the core of the star starts to shrink and get hotter, while the outer layers of hydrogen expand and cool, swallowing up any nearby planets in the process. At this stage in its life, a star is known as a **red giant**. Eventually, the central core becomes hot enough for the helium nuclei to fuse, forming carbon and oxygen. In small stars, as the helium in the core becomes depleted, the thermonuclear reactions will cease.

In more massive stars, a further collapse occurs, producing a temperature rise sufficient to trigger another round of thermonuclear reactions. The whole process is repeated, until a point is reached when no more energy can be released by further fusion reactions. By this stage, the centre of the core will consist of iron. Elements heavier than iron aren't formed, because this would use energy rather than release it. Each round of thermonuclear reactions takes place in shorter and shorter times at higher and higher temperatures, with the material in the core compressed to greater and greater densities.

Once the thermonuclear reactions cease, the star will collapse and 'die'. How this happens depends on the mass of the original star.

What happens next?

Small stars (with a mass similar to that of our Sun) contract, shrinking in volume until the gravitational forces are balanced by outward-acting forces associated with the electrons in the core. As the collapse occurs, potential energy is converted into thermal energy; the star becomes white hot, and its outer layers drift away into space, forming a **planetary nebula**. (Despite their name, planetary nebulas have nothing to do with planets!) The hot, dense star that remains is called a **white dwarf**. Our Sun will shrink to about the size of the Earth – one teaspoonful of it then having a mass of around 1000 kg! White dwarfs eventually cool and fade over millions of years into invisible **black dwarfs**.

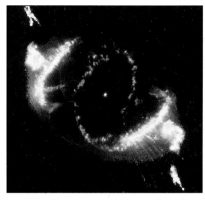

A planetary nebula with a white dwarf at its centre.

 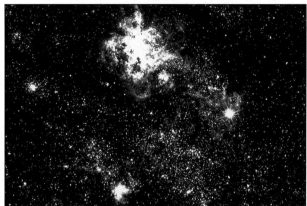

'Before' and 'after' photos. The right-hand photo shows the presence of a supernova. Can you spot it?

Larger stars collapse to a far greater extent because of the immense gravitational forces. In less than a second after thermonuclear reactions cease, the already dense material in the core is squeezed together until it is as densely packed as the material inside an atom's nucleus. The catastrophic implosion of the core causes material surrounding it to fall towards it at speeds at up to 15% of the speed of light. On meeting the now incompressible core, the outer material bounces off, and a massive explosion occurs, blowing the outer layers away into space and causing the star to shine as brightly as millions of suns. Superheating occurs, setting off a chain of thermonuclear reactions in which energy is absorbed. The heavier elements, including gold, uranium and lead, are formed and ejected – this is the source of these elements here on Earth. Such explosions are called **supernovas**. They are thought to occur about once a century in our galaxy.

The core remaining after a supernova is a small, dense **neutron star**, so called because it is believed to be composed entirely of neutrons. Neutron stars are only about 20–30 km in diameter, and one teaspoonful would have a mass of 10^{12} kg!

If the original star was sufficiently massive, the neutron star continues to shrink until it becomes a point of infinite density, whose immense gravity will allow nothing to escape – not even light – a **black hole**. Black holes are 'seen' by the effect of their huge gravitational field on other stars.

QUESTIONS

1. The supernova in the photo above occurred in a nearby galaxy 150 000 years ago. Why was it not seen from Earth until 1987?

2. Why does nuclear fusion eventually cease in a star?

3. At which points in a star's life are the following elements synthesised?
   a helium **c** iron
   b carbon **d** lead

11.7 The galaxies

The view from Earth

When we look at the sky at night with the naked eye, we can see at most a couple of thousand of the brighter and nearer stars in our own galaxy. Until the invention of telescopes, these were the only stars that astronomers were able to study – which they did by observing their positions and brightness. Even the nearest stars are so very far away that their relative positions seem scarcely to change when viewed from Earth, even over periods of thousands of years.

Telescopes have enabled observations of other galaxies to be made. It is estimated that the universe contains 50 to 200 thousand million galaxies scattered throughout space, each containing thousands of millions of stars.

Both elliptical and spiral galaxies can be seen in this photo.

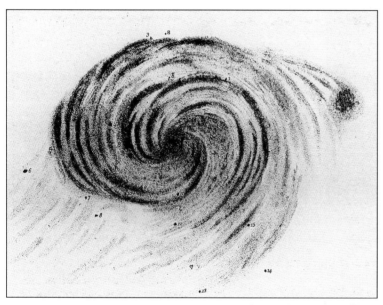

Lord Rosse's drawing (c. 1850) of the spiral galaxy now called M51.

Detailed observations of nearby galaxies have shown that there is a vast array of shapes and sizes. About 150 years ago the astronomer Lord Rosse studied nearby galaxies with a large optical telescope, and made drawings of their structure.

Details of the structure of our own galaxy, the Milky Way, didn't emerge until after the advent of radio telescopes earlier this century. Detection of radio emissions from interstellar gas has shown that, like many other galaxies, it has a spiral shape. There is a central elliptical region consisting mostly of old stars, surrounded by spiralling arms consisting of younger stars (including our Sun) and clouds of gas and dust where new stars are forming. The whole galaxy is rotating about its centre. Our Sun, travelling at a speed of about 250 000 m/s, will take about 200 million years to orbit the galactic centre.

Our place in the galaxy (an artist's impression).

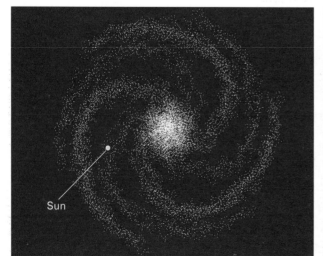

a Top view.

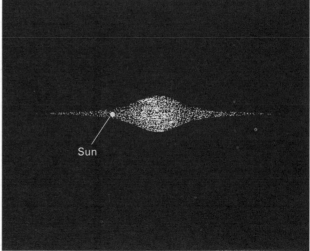

b Sectional view.

The galaxies are moving apart

The billions of galaxies in the universe are moving apart from each other at great speed. This was deduced by the American astronomer Edwin Hubble in the 1920s from an analysis of their light. Just as the pitch of a siren (whose frequency is constant) will appear to change from a higher pitch as it is moving towards you, to a lower pitch as it is moving away, so too will the frequency (colour) of light received from a fast-moving source.

Hubble found that, when the light from a distant galaxy was split into a spectrum of its constituent wavelengths, the colours were 'shifted' towards the red end. This effect is known as **red-shift**. It suggested that the galaxies are moving away from us – red light, like low-pitch sounds, has a longer wavelength. The further away the galaxy was, the greater the amount of red-shift Hubble observed.

The Big Bang

Hubble deduced that the further apart neighbouring galaxies are, the more rapidly they are separating. He developed the simple relationship that the speed of separation is proportional to the distance apart. The expansion was the same, no matter in which direction he looked. This led to the suggestion that the universe was originally concentrated into an ultra-dense point, that exploded some 15 thousand million years ago – the **Big Bang**.

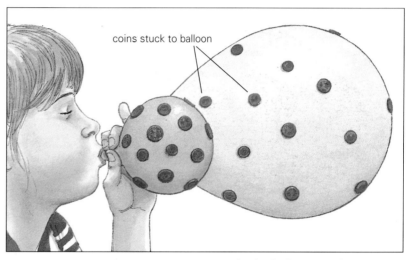

As the balloon is inflated, the attached coins will move apart in a way similar to the galaxies in the expanding universe.

An alternative theory to account for the red-shifts was that the universe is expanding but matter is continuously created throughout space, so that the structure and density of the universe remains constant. This is the 'steady-state' theory. It is no longer upheld by cosmologists because of recent evidence for the Big Bang theory. The most important discovery has been the 'cosmic background radiation' – microwave radiation detected in all directions which has no explanation other than being a remnant of the radiation from the Big Bang itself.

Recently, a space probe, the Cosmic Background Explorer, provided data which confirmed that the source of the background microwave radiation must have been the Big Bang.

QUESTIONS

1. Explain why, despite all the movement that is going on in space around us, the relative positions of stars do not appear to change.

2. Why do people think that the universe might have started with a Big Bang?

3. What might stop the expansion of the universe? How would observations of red-shift reveal a slowing-down of the expansion?

11.8 The evolution of our solar system

As our Sun was forming, the once shapeless nebula of gas and dust became transformed into a rotating flattened disc. While the temperature at the centre of the disc rose, the edges remained cold. Away from the centre, particles began to coalesce to form 'proto-planets', and eventually planets. When the temperature of the proto-Sun was high enough, thermonuclear reactions started. It is thought that both the Sun and the planets took about 100 million years to form.

The formation of the inner planets

Towards the centre of the disc of dust and gas, the four inner planets formed when solid rocky particles (with high melting points) coalesced. The temperature was such that most of the molecules of the 'lighter' gases (hydrogen and helium) had a sufficiently high velocity to prevent them from being trapped by the gravity of the forming planets. Over time, the particles coalesced into bigger and bigger lumps of material. The energy released as these large lumps (the proto-planets) collided with each other, combined with the energy released from radioactive elements within them, caused the rock to melt. The denser iron-rich minerals sank towards the centre, while the less dense silicon-rich minerals floated to the surface.

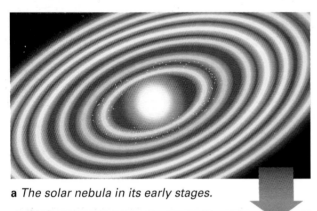

a *The solar nebula in its early stages.*

The formation of the gas giants

Meanwhile, towards the edge of the disc, rocky particles were also coalescing, forming the rocky cores of what were to become the planets Jupiter, Saturn, Uranus and Neptune. As it was much cooler here than near the centre of the disc, the surrounding molecules of hydrogen and helium had much lower average velocities and were gradually captured, until the planets were sufficiently large for their gravity to pull in all the remaining gas that was available.

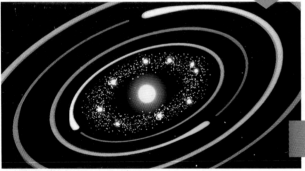

b *The solar system after 50 million years.*

c *After 100 million years the Sun and planets were much as they are today.*

Asteroids and comets

Between the inner and outer planets, many millions of lumps of rocky material which didn't coalesce further are in orbit. They are called **asteroids**, and make up the 'asteroid belt'. They have irregular shapes, because their mass is insufficient for their own gravity to have pulled them into the spherical shape of a planet. Their dimensions vary, but can be as much as 1000 km. The asteroid belt contains the majority of the asteroids in our solar system.

Towards the edge of the solar system, the uncoalesced material (like the outer planets) become covered with substantial amounts of frozen gases. Some of these lumps have highly elliptical orbits which take them into the inner solar system. When they pass near the Sun, some of the gases are vaporised and we see a **comet**.

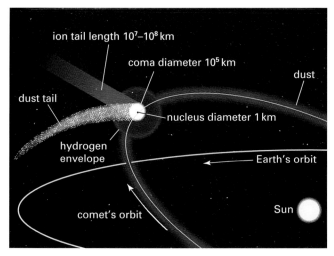

A typical comet when close to the Sun – seen as a bright head (the coma), with a tail streaming millions of kilometres into space (in a direction directly away from the Sun).

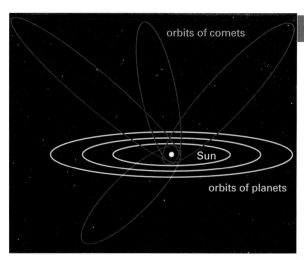

The orbits of comets are not in the same plane as most of the rest of the solar system.

Most comets are invisible to the naked eye but occasionally there is a spectacular sighting. One of the most famous is Halley's comet (named after the second Astronomer Royal), which orbits the sun every 76 years. Every arrival in the inner solar system since 87 BC has been recorded by astronomers. It was investigated by the Giotto space probe when it last arrived in 1986.

The appearance of Halley's comet in 1066 is recorded in the Bayeux tapestry.

Collisions with the planets

The appearance of the surface of the inner planets suggests that collisions with rocks and other debris is not an uncommon experience. Some people speculate that the unusual orientation of the rotation axis of Uranus (almost parallel to the plane of its orbit) could have been caused by it being 'knocked on its side' by a planet-like object. It has been estimated that 300 000 kg of material enters the Earth's atmosphere every day. However, as it is slowed by frictional forces, it gets so hot that most of it is burnt up before it reaches the Earth's surface. Some craters on the Earth's surface, however, are the result of collisions with massive lumps of rock. Some people argue that the dinosaurs became extinct as the result of such a collision. Fortunately, collisions with objects large enough to wipe us out are extremely unlikely to occur during our lifetimes.

This impact crater in the Arizona desert has a diameter of over 1 km.

Exploring the solar system today

In recent years, many unmanned space probes have been sent out into the solar system to take photographs and collect information about the planets and their moons. Amongst the most famous were Voyager 1 and Voyager 2, which were launched in 1977. Such probes use the gravity of the planets to accelerate them on their way and have to be carefully programmed to follow the desired paths.

QUESTIONS

1. Why is the Earth's surface rich in silicon minerals?
2. Why do most of the planets in our solar system orbit in the same plane?
3. Explain why a comet can only be seen at certain points in its orbit.

11.9 Moving round in circles

In everyday life, as in space, there are many examples of bodies moving in paths which, if not exactly circular, are nearly so. Before looking at why objects tend to move in this way in space, it is helpful to study a more familiar example.

When a conker is swung around on the end of a string in a circle at a steady speed, its velocity is continuously changing because its direction of travel is constantly changing (see 4.3). This means that it must be accelerating; and for it to accelerate, there must be a resultant force acting (see 4.5).

Where does the resultant force come from?
There are two forces acting on the conker: its weight, and the tension in the string. When these are added together, their resultant is a horizontal force. The direction of the resultant changes as the conker moves around the circle, but it is always directed along a radius and towards the centre of the circle. The acceleration it produces is also therefore towards the centre.

The size of the force towards the centre to keep an object of mass m moving with a velocity v in a circle of radius r is given by the formula:

$$\text{force (N)} = \frac{\text{mass (kg)} \times \text{velocity}^2 \, [(\text{m/s})^2]}{\text{radius (m)}}$$

or, in symbols:

$$F = \frac{mv^2}{r}$$

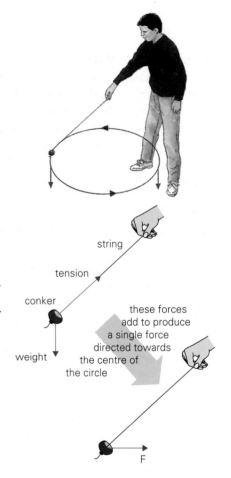

It is the forces of gravity between the planets and the Sun, and between the Earth and its satellites, which are holding each of them in orbit. The force on an orbiting body is always directed along a radius towards the centre of the orbit, just as in the case of the conker.

How big is the force of gravity?
There are gravitational forces of attraction acting between all objects. The size of the force depends on the mass of each of the objects, and their separation. The force on each object is equal and opposite, and is given by:

$$\text{gravitational force (N)} = \frac{G \times \text{mass of object 1(kg)} \times \text{mass of object 2(kg)}}{\text{separation}^2(\text{m})^2}$$

or, in symbols:

$$F = \frac{Gm_1 m_2}{r^2}$$

where G is the gravitational constant. It has a value of 6.7×10^{-11} Nm^2/kg^2.

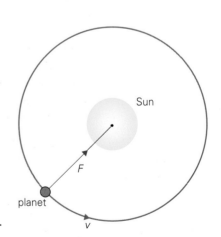

Doubling the mass of either object doubles the size of the gravitational force, whereas doubling their separation reduces the size of the force to a quarter of its former magnitude.

The gravitational field strength of the Earth, g, is the gravitational force on unit mass (1 kg). Therefore:

$$g = \frac{Gm_E}{r^2}$$

where m_E is the mass of the Earth and r here is the distance *from the centre of the Earth*. This shows that the gravitational field strength decreases with height above the Earth's surface, a fact which is very important in calculating the orbits of satellites.

Satellites in orbit

In order to remain in orbit at a particular height above the Earth, a satellite needs to have a particular velocity. It can be shown (from the formulas opposite) that for different objects (A and B) in orbit around the same central mass (e.g. the Earth) then, irrespective of the objects' masses:

$$v_A^2 r_A = v_B^2 r_B$$

where v represents orbital velocity and r the radius of the orbit. It follows that the higher a satellite, the larger the value of r, and so the lower its orbital velocity. Similarly, the further a planet is from the Sun, the lower its velocity. This, together with the fact that they have to travel further, explains the very long orbit times of the outer planets (see 11.1).

In a **geostationary** or **geosynchronous orbit**, the satellite orbits the Earth once every day. Since the Earth turns on its axis once every day, this means that the satellite will always be positioned above the same point on the Earth's surface (somewhere on the equator). Such satellites are used for communications. Information in the form of microwave signals can be transmitted to them, and received from them in another part of the world. BSkyB television, for example, is transmitted from an Astra satellite. All satellite dishes are pointed towards the same place in the sky to receive broadcasts. Geostationary satellites are also used for long-distance telephone links, and by ships at sea to determine their exact position.

An artist's impression of the Astra 1b satellite in orbit.

The closer a satellite is to the Earth, the faster it moves, and hence the shorter the time it takes to orbit the Earth once. Low satellites are used to 'spy' on the Earth, sending information back to receivers which track them (the angle of the receiver needs to be continually adjusted, unlike the receivers used with geostationary satellites). These observation satellites are normally put into a **polar orbit** (so that they pass over both poles). They are able to scan the whole of the Earth's surface as it rotates on its axis beneath them. They are used for monitoring the weather and changes to the Earth's surface, as well as for spying on other countries.

Very recently a new type of satellite has been put into use. The Hubble Space Telescope was launched in 1990, and since 1993 has been transmitting pictures of distant galaxies. They are much clearer than images from ground-based telescopes because these suffer distortion due to the Earth's atmosphere. The image of the planetary nebula in 11.6 came from this source.

QUESTIONS

1. What is the force which makes the planets orbit the Sun? In which direction does it act?

2. What factors do all geostationary satellites located above the equator have in common?

3. Why aren't communications satellites put into polar orbits?

SECTION F: QUESTIONS

For these questions take the velocity of light as 3×10^8 m/s.

For other data refer to the table on 11.1

1a Draw a diagram to explain how the orbits of comets differ from the orbits of planets.
b What makes a comet visible?
c Explain why a planet in a circular orbit around the Sun will have a constant speed, but not a constant velocity.
d Why doesn't a comet travel at a constant speed as it orbits the Sun?

2 List the planets in order of:
a their distance from the Sun,
b their diameter.

3 About how long does light from the Sun take to reach the planet Pluto? Why can't you give a precise answer?

4 Explain why it is still possible to see a star that no longer exists.

5

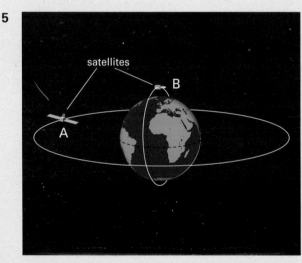

a Which of the two satellites in the diagram:
  i is orbiting the Earth the fastest?
  ii could be in a geostationary orbit?
  iii is in a polar orbit?
b If satellite B were in a higher orbit, what effect would this have on the time taken to orbit the Earth?
c Name two uses of satellites:
  i in geostationary orbits
  ii in polar orbits.
d A force is needed to keep a satellite in orbit.
  i In what direction does the force act?
  ii What does the size of the force depend on?

6a What is 'red-shift'?
b Why did Hubble deduce that the universe was expanding?
c How did Hubble suggest that the universe might have started?

7a Why are we able to see
  i stars?
  ii planets?
b Suggest two different factors that will affect how bright a star appears.

8a Explain:
  i how stars are believed to begin their lives,
  ii how a young star gets hot enough to start the fusion reactions in its core.
b How does a star like our Sun differ from a white dwarf?
c What is a supernova?
d Why are neutron stars so dense?
e At what stage in a star's life are elements heavier than iron formed?

9 Place the following in order of size:
galaxy, red giant, Jupiter, our Sun, the universe, our solar system, asteroid, white dwarf.

10 Explain, with the help of a diagram, why, as the planets move across the constellations of the zodiac, their paths appear to loop back on themselves.

11 What evidence emerged in the eighteenth century to confirm the view that the Earth was in motion around the Sun?

12 Explain why the appearance of the Moon changes on a regular basis.

13 Each day, the Earth moves about 1° around its orbit of the Sun.
a How long does the Earth take to rotate through 1° on its axis?
b Explain with the help of a diagram why each day on the Earth is about 4 minutes longer than the time it takes for the Earth to rotate through 360° on its axis.
c How many times will the Earth rotate on its axis in the time that it takes to orbit the Sun once?

14a Copy the diagram.

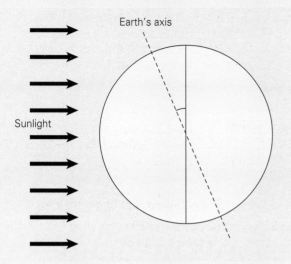

  i Mark the region of the Earth where it is day, and the region where it is night.
  ii Mark the region of the Earth which will be experiencing continuous daylight (i.e. where it won't get dark at night).

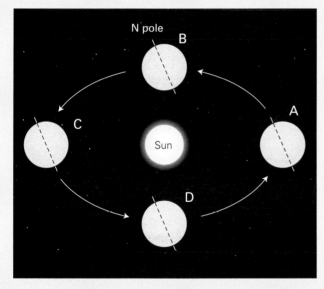

b At which points A, B, C, D in the Earth's orbit will people in the southern hemisphere experience
  i summer,
  ii winter,
  iii spring,
  iv autumn?
c Using diagrams, explain why there is more daylight in the summer than there is in the winter.
d Explain why it is colder in the winter than in the summer.

15a Assuming that the orbits of the Earth and Jupiter are circular and in the same plane, calculate the distance between the two planets when their separation is at:
  i its maximum,
  ii its minimum.
b If light from Jupiter could travel directly to the Earth (i.e. if the Sun wasn't in the way), calculate how long it would take the light to arrive at the Earth when their separation is at:
  i its maximum,
  ii its minimum.
c By what factor will the gravitational force between the two planets increase as their separation changes from its maximum to its minimum?

16 The Sun has a diameter of 1.39×10^6 km. A student making a scale model of the solar system chose to use a ball of diameter 30 cm to represent the Sun. Calculate:
a the diameter of each of the planets in the model,
b the distance they should be placed from the Sun.

17 How were astronomers able to deduce the presence of the planet Neptune without actually having observed it?

18a Why is there so much hydrogen present in Jupiter's atmosphere, but scarcely any in the Earth's?
b Why didn't Jupiter develop into a star?

19 If, like Mercury, the Earth's axis was tilted at an angle of just 22° rather than at 23.5°, what effect would this have on:
a the amount of daylight on any particular day,
b the seasons,
c the way in which trees have evolved?

20 While there are many impact craters visible on the Moon, there are very few visible on the Earth. Suggest three reasons why.

21 If the distance from the Earth to the Moon were to decrease, what effect would this have on the time that the Moon takes to orbit the Earth?

22 Why would the pattern of stars within a constellation change if it were viewed from different points within the galaxy?

Index

A

a.c. *see* alternating current
acceleration 48–51
    falling objects 50–51
active solar heating 124
activity (rate of radioactive decay) 134
Adams, John 147
alpha rays (particles) 131, 132
    biological effects 138
    in smoke alarms 140
alternating current 7, 20
    generation 34, 35, 38
    rectification 36
alternators 34, 38
aluminium, extraction 16
aluminium alloys 60
amber 2
americium-241, in smoke alarms 140
ammeter 7, 8, 10
ampere (unit) 6, 29
Ampère, André Marie 29
amplification (sound) 93
amplitude 71
    sound waves 93
aneroid barometers 67
angle of incidence 80, 82, 84
angle of reflection 80, 82, 84
anode 16
asteroids 160
atmosphere (unit) 67
atmospheric pressure 63, 66–7
atomic bombs 142
atomic number 130
atoms 3, 56
    structure 130
audible range of sound 92
average speed 46

B

babies, ultrasound scanning 95
background radiation 138–9
    cosmic 159
barometers 67
bats, echo-location 94
batteries 6, 7, 20, 103
becquerel (unit) 134
Becquerel, Henri 130
bell (electric) 30
beta rays (particles) 131, 132
    biological effects 138
    in thickness gauges 140
Big Bang 159
binoculars 85
black dwarfs (stars) 157
black holes 157
'Bogman' 136
bonds 90
Boyle, Robert 66
Boyle's Law 66
Bradley, James 153
braking systems 65
Brownian motion 57
brushes (in electric motor) 32
building materials 60

C

cancer
    of skin 77
    treatment with gamma rays 79, 141
    treatment with X-rays 4, 78
carbon-14, decay 131
carbon dating 136
carbon dioxide 118
carbon resistors 10
cars
    acceleration 49
    brakes and tyres 52
    braking systems 65
    forces on 45
    lighting circuits 13
    spark circuit 35
    starter motors 31
    stopping distance 46–7
    streamlining 52
cat, falling 53
cathode 16
cathode ray oscilloscope (CRO) 17, 20, 92–3
cavity wall insulation 106
CD player 24
cells (electric) 6, 7
    connecting 12
chain reaction (nuclear) 142
chemical energy 100, 101
Chernobyl accident 119
chimneys 107
chloro-fluoro-carbons 77
circuit breakers 22
circuit diagrams 6–7, 14
circuits 6
    materials for 19
    series and parallel 8, 12–13, 14
    using mains supply 21
circular motion 51, 162–3
cleaning, using ultrasound 95
cobalt-60, use in radiotherapy 141
colours (light) 83
combined cycle gas turbines (CCGTs) 118, 126
combined heat and power stations 126
comets 160–61
communications satellites 32, 75, 163
commutator 32, 33
components of circuits 6
compression 59
conduction, thermal 104, 106
conductors, electrical 3
    non-ohmic 10
    ohmic 9, 10
constellations 148
consumer unit 22
convection 105
convection currents 61, 105, 107
cooling fins 109
Copernicus, Nicolaus 152
core, Earth's 97
Cosmic Background Explorer 159
cosmic background radiation 159
cosmic rays 138
coulomb (unit) 14
Coulomb, Charles Augustin de 14
count rate 133
critical angle 84
CRO *see* cathode ray oscilloscope
crust, Earth's 97
Curie, Marie 130
current *see* electric current

D

d.c. *see* direct current
day length 150
decay chains 137
deceleration 48
density 60–61
depth-testing 94
deviation of ray 82, 83
diffraction 86–7
diffuse reflection 80
digital coding 84
diode 7, 11
direct current 20, 36
    generation 34
dispersion 83
distance–time graphs 46
district heating schemes 122, 123
dog whistles 94
domestic appliances (electrical) 24–5
    cost of running 26

dose (radiation) 138, 139
double insulation 23, 25
drag 52–3
dragsters 49
draught excluder 111
dynamo 6, 34

E

Earth 146
   age 137
   structure 97
   tilt of axis 150
earth wire 22, 23
earthing 4, 22
earthquakes 96–7
echo-location 94
echo-sounding 94
echoes 89, 91
eddy currents 37
effective dose (radiation) 138
efficiency 115
   electrical devices 103
   power stations 126
Einstein, Albert 142
elastic limit 58
elastic potential energy 100
elasticity 58–9
electric bell 30
electric charge 2–3
   movement *see* electric current
electric current 3, 6, 14–15
   induced 34
   magnetic effect 29
   measuring 8
   and Ohm's Law 8–9
electric motors 32–3, 34
electric plugs 23, 25
electric shocks 3
   protection from 23
electrical energy 100
electrical heating 13, 18–19, 27
electrical power 18–19
electricity 2
   cost of 26, 126
   energy sources for production 117
   generation 34–5, 38, 118–21
   matching supply and demand 127
   off-peak 27
   transmission 38–9
electricity meter 22, 26
electrodes 16
electrolysis 16
electrolyte 16
electromagnetic induction 34–5

electromagnetic radiation 74, 100
electromagnetic spectrum 74–5
electromagnetic waves 74–9
electromagnets 29–31, 34
electrons 3, 14, 130
electrostatic forces 2
endoscope 85
energy
   conversions 100, 101, 102–3
   in electrical circuit 15
   forms 100
   related to mass 142
   saving in home 110–11
   sources 101, 116–17
environmental damage/costs 117
epicentre of earthquake 96, 97
equilibrium 54
equinoxes 150
equivalent dose (radiation) 138
ethanol production 125
evaporation 105
extension–load graphs 58

F

falling objects 50–51, 53
   speed 113
fan heater 13
Faraday, Michael 34
fermentation 125
fibre optics 84–5
field lines 28, 29
filament lamp 10, 24
film badges 132, 133
fire-fighting suits 109
flat roofs 109
flaw detection 95
Fleming's left-hand rule 32
floating 61
flow detectors 141
flue-ash precipitation 5, 118
fluorescent lighting 77
foams, insulating 106
focus of earthquake 96, 97
force magnifiers 64
forces 42–3
   and acceleration 51–2
   equal and opposite 43
   resultant 44–5
   work done by 114
fossil fuels 101, 116
   combustion products 118
Franklin, Benjamin 2, 4, 14
frequency 71, 72
   mains electricity supply 20, 34, 35

sound waves 92
friction 52–3
fuels 100, 101
fuses 7, 22, 23, 25
fusion energy 154

G

galaxies 154, 158–9
Galileo Galilei 153
gamma rays 78–9, 131, 132
   biological effects 138
   use in radiotherapy 79, 141
gases 56
   compressing 57
   pressure in 66–7
   speed of sound in 90
'gasohol' 125
Geiger–Müller tube (counter) 133
generators (electrical) 6, 34–5, 118
geostationary (geosynchronous)
   orbit 163
geothermal energy 122–3
global warming 118, 125
gravitational constant 162
gravitational field strength
   42, 50, 162
gravitational force 42, 162
gravitational potential energy
   100, 112, 113
greenhouse effect 125
greenhouse gases 118, 123, 125
grip 52
Guericke, Otto von 67

H

hairdryer 25
half-life 134–5
Halley's comet 161
heart surgery 85
heat 104
heat pumps 110
heating
   electrical 13, 18–19, 27
   solar 124, 125
heating elements 24–5
Herschel, William 147
hertz (unit) 71
Hooke, Robert 58
Hooke's Law 58
horsepower (unit) 115
hot-water system 107
household wiring 22–3
houses, energy saving in 110–11
Hubble, Edwin 159

167

Hubble Space Telescope 163
hydraulic braking systems 65
hydraulic jacks 65
hydraulic systems 64–5
hydroelectric power stations 120
hydrogen, nuclear fusion 154

I

Iceland, geothermal energy 123
igneous rocks, dating 137
images 80, 81
immersion heater 25, 107
induced voltage 34
Information Super-Highway 84
infra-red radiation 108
infra-red waves 76, 79
insulators
    electrical 3
    thermal 104, 106
internal energy 100
    see also thermal energy
ionic compounds 16
ionising radiation
    background 138–9
    biological effects 138
    doses 138, 139
ions 16
iridium-183, in flow detectors 141
iron (electric) 24
isotopes 131

J

Joint European Torus 155
joule (unit) 15, 64, 100, 114
Joule, James 100
Jupiter (planet) 146, 148, 149, 153

K

'keyhole' surgery 85
kidney stones, treatment 95
kilogram (unit) 42
kilohertz (unit) 71
kilojoule (unit) 18
kilowatt (unit) 18, 115
kilowatt-hour (unit) 26
kinetic energy 100, 112, 113
kinetic theory of matter 57

L

lamps 7, 10, 24
    fluorescent 77
lasers
    in CD player 24
    diffraction of beam 86
    in optical communications 84
lateral inversion of image 81
le Verrier, Urbain 147
levers 54
light
    diffraction 86
    reflection 80–81
    refraction 72, 82–3
    transmission in optical fibres 84–5
    waves 80–81
light bulbs 102
    efficiency 103
light-dependent resistors (LDRs) 11
light-emitting diodes (LEDs) 76
light-year (unit) 154
lighting circuits 13, 22
lightning 4
lines of force 28, 29
liquids 56
    compressing 57
    immiscible 61
    pressure in 63
live wire 21, 23
longitudinal waves 70, 71, 88
loudness 93
loudspeakers 33, 93
lubricants 52

M

Maglev train 30
magnetic fields 28–9, 32, 33
magnets 28
    permanent 30
    see also electromagnets
mains adaptors 36
mains electricity 20–21
mantle, Earth's 97
Mars (planet) 146, 149, 152
mass 42
    equivalent to energy 142
mass number 130
mechanical energy 100
medical tracers 79, 141
megahertz (unit) 71
megawatt (unit) 115
Mercury (planet) 146, 148
metals
    deposited in electrolysis 16
    electrical conductors 3, 10
    thermal conductors 104, 106
methane, produced from waste 123
microphone 35
microsievert (unit) 138
microwaves 75, 79
diffraction 87
Milky Way 154, 158
mirrors 80, 81
molecular theory of elasticity 59
molecules 56
    movement 57
moment 54
Moon 151
    acceleration of falling object on 51

N

National Grid 38, 39
natural gas 118
nebulae 156, 157
Neptune (planet) 146, 147
neutral wire 21, 23
neutron stars 157
neutrons 3, 130
newton (unit) 42, 50
Newton, Isaac 42
Newton's laws of motion 43, 45
nitrogen oxides 118
noise 88
noise pollution 93
non-ohmic conductors 10
non-renewable energy sources 116, 117
normal to mirror 80
nuclear accidents 119
nuclear energy 100, 142–3
nuclear equations 131
nuclear fission 142
nuclear fusion, in stars 154, 155
nuclear power stations 117, 119, 126, 143
nuclear radiation 130, 131, 132–3
nuclear reactors 143
nuclear waste, disposal 143
nucleon number 130

O

Ocean Thermal Energy Conversion plant 121
off-peak electricity 27
ohm (unit) 8
Ohm, Georg 8
Ohm's Law 8–10
ohmic conductors 9, 10
optical communication 84–5
optical fibres 84–5
optical instruments 85
oscillations 70, 72
Osprey 1 wave generator 121
ozone layer 77

P

P waves 96, 97
painting, spray- 5
parallel circuits 8, 12–13, 14
pascal (unit) 62
Pascal, Blaise 63
pendulum 112
period of oscillations 72
permanent magnets 30
phases of Moon 151
photocopiers 5
photosynthesis 101, 125
pitch 92
planetary nebulae 157
planets 146–9
    apparent motion 152
    collisions with 161
    formation 160
plasma 155
plugs (electric) 23, 25
Pluto (planet) 146, 147, 149
polar orbit 163
poles of magnet 28
pollution 117
polonium-210, decay 131
potassium:argon dating 137
potential difference 4, 6
potential energy 100, 112
power 115
    electrical 18–19
power stations 118–19
    efficiency 126
    flue-ash precipitation in 5
    nuclear 117, 119, 126, 143
pressure 62–3
    in gases 66–7
    in liquids 63
pressure waves 92
primary energy sources 117
prisms 82–3, 85
proton number 130
protons 3, 130
protostars 156
Ptolemy 152
pumped storage hydroelectric schemes 120, 127

R

radiation 108–9
radio waves 74, 79
    diffraction 87
radioactive decay 130, 134–5
radioactive decay chains 137
radioactivity 130–31

radiocarbon dating 136
radioisotopes 131
    decay 134–5
    half-lives 134–5
    storage 132
    uses 140–41
radium-221, half-life 134
radon 138–9
radon-220, half-life 135
rays of light 80
RCDs (residual current devices) 23
rectification 36
red giants (stars) 156
red-shifts 159
reflection 72, 80–81
    total internal 84–5
refraction 73
    light 82–3
    sound waves 72, 90
refrigerator 24
regular reflection 80
relays 31
remote controls 76
renewable energy sources 116, 117
residual current devices (RCDs) 23
resistance 8–10
    changes with temperature 10, 11
    in a circuit 12
    and heating effects 18–19
resistors 8
    light-dependent 11
    variable 7
resultant force 44–5
reverberation 89
right-hand grip rule 29
right-hand screw rule 29
ring main 22
ripple tank 72–3
road safety 46–7
robots 31
Röntgen, Wilhelm 78
Rosse, Lord 158
Rutherford, Ernest 130

S

S waves 96, 97
satellites 32, 75, 76
    orbits 163
Saturn (planet) 146
scalar quantities 44
screwdriver 55
seasons 150
seismic waves 96–7
seismograms 96, 97

seismometer 96
semiconductor components 11
series circuits 8, 12–13, 14
sidereal day 146
sievert (unit) 138
sinking 62
skin, damage by ultra-violet waves 77
slip rings 35
smoke alarms 140
solar cells 124
solar day 146
solar energy 100, 101, 125–6
solar furnaces 124
solar heating 124, 125
solar panels 124
solar system 146–7, 158
    evolution 160–61
    exploring 161
solenoids, magnetic field 29
solids 56
    compressing 57
    speed of sound in 90
solstices 150
sonar 95
sound energy 100
sound waves 70
    diffraction 87
    high-frequency (ultrasound) 94–5
    picturing 92–3
    producing 88–9
    reflection (echoes) 89
    refraction 72, 90
    speed 90–91
'space blankets' 109
space probes 159, 161
sparks (electric) 4
spectrum
    electromagnetic 74–5
    visible light 83
speed 46–7
spray-painting 5
spring balance 43, 45, 58
springs
    stretching 58
    waves on 70, 71
stars 148
    evolution 156–7
static electricity 2
    hazards 4
    uses 5
stellar aberration 153
sterilisation, using gamma rays 79
storage heater 27
streamlining 52

169

'suction pads' 66
sulphur dioxide 118
Sun 154–5, 156
   source of energy 101, 155
sunbeds 77
sunburn 77
supernovas 157
switches 7
symbols
   circuit diagrams 6, 7
   isotopes 131

T

tape recordings 35
technetium-99, use as tracer 141
tectonic plates 96
telecommunications 75
telescopes 153
television set
   remote control 76
   transformers in 36
temperature 104
   effects on density 61
   effects on pressure 66
   effects on resistance 10, 11
tennis racquet and ball 43
tension 59
terminal velocity 53
thermal conductivity 104, 106
thermal energy 100, 102, 103
   extraction from sea 121
   transferring 104–5
thermals 107
thermistor 11
thermonuclear reactions, in stars 154, 156, 157, 160
thickness gauges 140
tidal power stations 119, 120
time period of oscillations 72
torque 54
torque wrench 55
torus 155
total internal reflection 84–5
tracer isotopes 79, 141
transformers 36–7, 38–9
transverse waves 70, 71
tuning fork 88
turbines 118
Turin Shroud 136
turning effects 54–5

U

U-values 110
ultrasound (ultrasonic waves) 94–5
ultra-violet waves 77, 79
units
   acceleration 48
   density 60
   distance 154
   electric charge 14
   electric current 6, 29
   electrical resistance 8
   electricity used 26
   energy 15, 18, 100
   force 42
   frequency 71
   mass 42
   power 18, 115
   pressure 62, 63
   radiation dose 138
   radioactivity 134
   speed 46
   velocity 47
   voltage (potential difference) 6
   work 64, 114
uranium-235, decay 142, 143
uranium-238 decay series 137, 138
Uranus (planet) 146, 161
useful work 115

V

vacuum flask 109
variable resistor 7
vector quantities 44
velocity 47
   terminal 53
velocity–time graphs 48
Venus (planet) 146, 148
virtual images 81
visible light 79, 80
volt (unit) 6
Volta, Alessandro 6
voltage 4, 6, 15
   induced 34
   measuring 8
   and Ohm's Law 8–9
voltmeter 7, 8, 10
Voyager space probes 161

W

washing machine motor 33
waste, energy from 123
water
   heating 106
   variations in density 61
water-heating system 107
water waves 70, 72
   diffraction 86
   reflection 72
   refraction 73
watt (unit) 18, 115
Watt, James 115
wave power 121
wavelength 70, 71
   sound waves 92
   water waves 72, 73
waves 70–71
   electromagnetic 74–9
   in ripple tank 72–3
   speed 71
   types 70
weight 42, 50
whale song 88
wheelbarrow 55
white dwarfs (stars) 157
wind 107
wind power 122
work 114

X

X-rays 78, 79

Z

zodiac 148